NANOTECHNOLOGY AND ITS APPLICATIONS

To learn more about AIP Conference Proceedings, including the Conference Proceedings Series, please visit the webpage
http://proceedings.aip.org/proceedings

NANOTECHNOLOGY AND ITS APPLICATIONS

First Sharjah International Conference on
Nanotechnology and Its Applications

*Jointly organized by the American University of Sharjah
and the University of Sharjah*

Sharjah, United Arab Emirates   *10 – 12 April 2007*

EDITORS

Y. I. Salamin
N. M. Hamdan
*American University of Sharjah
Sharjah, United Arab Emirates*

H. Al-Awadhi
N. M. Jisrawi
*University of Sharjah
Sharjah, United Arab Emirates*

N. Tabet
*KFUPM
Dhahran, Saudi Arabia*

All papers have been peer reviewed.

SPONSORING ORGANIZATIONS
Al-Sharif Group, L.L.C.
ADNOC and Its Group of Companies
Petrofac
Sharjah Chamber of Commerce and Industry
Sharjah Municipality
Dubai Silicon Oasis Authority
Veeco
Agilent Technologies

Melville, New York, 2007
AIP CONFERENCE PROCEEDINGS ■ 929

P1475

Editors

Y. I. Salamin
N. M. Hamdan

Physics Department
American University of Sharjah
P.O.B. 26666
Sharjah, United Arab Emirates
E-mail: ysalamin@aus.edu
          nhamdan@aus.edu

H. Al-Awadhi
N. M. Jisrawi

Physics Department
University of Sharjah
Sharjah, United Arab Emirates

N. Tabet

Physics Department
KFUPM
Dhahran, Saudi Arabia

Authorization to photocopy items for internal or personal use, beyond the free copying permitted under the 1978 U.S. Copyright Law (see statement below), is granted by the American Institute of Physics for users registered with the Copyright Clearance Center (CCC) Transactional Reporting Service, provided that the base fee of $23.00 per copy is paid directly to CCC, 222 Rosewood Drive, Danvers, MA 01923, USA. For those organizations that have been granted a photocopy license by CCC, a separate system of payment has been arranged. The fee code for users of the Transactional Reporting Services is: 978-0-7354-0439-7/07/$23.00

© 2007 American Institute of Physics

Permission is granted to quote from the AIP Conference Proceedings with the customary acknowledgment of the source. Republication of an article or portions thereof (e.g., extensive excerpts, figures, tables, etc.) in original form or in translation, as well as other types of reuse (e.g., in course packs) require formal permission from AIP and may be subject to fees. As a courtesy, the author of the original proceedings article should be informed of any request for republication/reuse. Permission may be obtained online using Rightslink. Locate the article online at http://proceedings.aip.org, then simply click on the Rightslink icon/"Permission for Reuse" link found in the article abstract. You may also address requests to: AIP Office of Rights and Permissions, Suite 1NO1, 2 Huntington Quadrangle, Melville, NY 11747-4502, USA; Fax: 516-576-2450; Tel.: 516-576-2268; E-mail: rights@aip.org.

L.C. Catalog Card No. 2007933083
ISBN 978-0-7354-0439-7
ISSN 0094-243X

Printed in the United States of America

CONTENTS

T
174
.7
I 53
2007
PHYS

THEORY

MODELING AND COMPUTATIONAL

QUANTUM WELLS, QUANTUM WIRES, AND QUANTUM DOTS

CARBON NANOTUBES

SYNTHESIS AND CHARACTERIZATION

BIO-NANOTECHNOLOGY

ELECTRONIC STRUCTURE

APPLICATIONS

MISCELLANEOUS

Preface

The *First Sharjah International Conference on Nanotechnology and Its Applications* (Sharjah-Nano07) was held during the period 10-12 April 2007 at the American University of Sharjah (AUS) in the United Arab Emirates. The conference was co-organized with the University of Sharjah.

This Conference brought together leading nano-science experts, academics, researchers, scientists, and students from various institutions around the world. It was an opportunity for scientists in the Gulf region and the Arab World to interact, exchange ideas and possibly start collaborations with colleagues in the region and in the international nano-community.

Such activities and meetings provide a platform where national and regional decision makers in charge of industries with critical importance to the Gulf region can meet with experts and discuss the impact of nanotechnology on their industries. This includes, but is not limited to, energy, food and agriculture.

The conference also turned the attention to the role that could be played by nanotechnology in dealing with environmental effects of industrial development in the region and in the world.

The conference had five main themes:

1. Environmental applications: sensors of pollutants, pollution remediation and green nanotechnology.
2. Petrochemical applications: petroleum refinement and production of gasoline from methanol.
3. Applications in the food industries and in agriculture.
4. Material properties and basic science aspects of nanotechnology such as synthesis, characterization, assembly and processing of nanostructures.
5. Theory, modeling and simulation.

The conference program included talks by seven internationally renowned keynote speakers, who delivered talks on various aspects of nanotechnology related to the above themes. In addition to that we received more than 220 abstracts, from which 141 were accepted, representing 64% of the total submission. Among the accepted papers 38 were set for oral presentations and 103 were presented in two poster sessions.

The program, furthermore, included a roundtable discussion that summarized the conference outcomes and attempted to lay out the ground work for a future roadmap for a nanotechnology initiative for Sharjah. Even though this conference had regional emphasis, with a significant number of papers submitted from institutions in the United Arab Emirates, it was truly international with about 30 other counties represented.

The proceedings contain peer-reviewed papers presented at the conference.

Following the conference, steps have been taken to establish a nanotechnology research center in Sharjah and several regional and international collaborations have already started. The second Sharjah-nano conference is planned for 2009.

We would like to thank our sponsors, the invited Speakers, the international advisory committee, the referees, and the participants. It has been a pleasure working with them all.

The Editors

The Organizing Committee

Chair:

Nasser M. Hamdan
American University of Sharjah, UAE

Co-Chairs:

Hussain Al-Awadhi
University of Sharjah, UAE

Nouar Tabet
King Fahd University of Petroleum and Minerals, Saudi Arabia

Members:

Najeh Jisrawi
University of Sharjah, UAE

Yousef Salamin
American University of Sharjah, UAE

Abdelrahman Al – Ali
American University of Sharjah, UAE

Mazhar Iqbal
American University of Sharjah, UAE

Wafaa Al – Haj
University of Sharjah, UAE

The Scientific Advisory Committee

- **Al – Najjar,** Arab Science and Technology Foundation (UAE).
- **C. Bai,** Chinese Academy of Sciences, and the National Center for Nano-science and Technology (China).
- **K. E. Geckeler,** Gwangju Institute of Science & Technology (South Korea).
- **Y. Haik,** United Arab Emirates University (UAE).
- **T. E. Madey,** Rutgers, the State University of New Jersey (USA).
- **W. I. Milne,** Cambridge University (UK).
- **M. Nayfeh,** University of Illinois at Urbana-Champaign (USA).
- **M. Prato,** University of Trieste (Italy).
- **T. Sekiguchi,** National Institute for Materials Science (Japan).
- **I. Siddiqi,** University of California at Berkeley (USA).

The Sponsors

Al-Sharif Group L. L. C.

ADNOC & Its Group of Companies

Petrofac

Sharjah Chamber of Commerce & Industry

Sharjah Municipality

Dubai Silicon Oasis Authority

Veeco

Agilent Technologies

Invited Talks

Nanoscience and Nanotechnology at Megafacilities
Helmut Dosch
Max-Planck Institute for Metal Research in Stuttgart, Germany

Milk Nanotubes and Self-Assembled Structures in Food Applications
C. G. de Kruif
Van 'T Hoff Laboratory of Utrecht University (the Netherlands) and Research Strategist and Senior Consultant at NIZO Food Research

Progress of Nanoscience and Nanotechnology in China
Chunli Bai
Executive Vice-president of the Chinese Academy of Sciences (CAS), Director of Division of Chemistry and Member of Executive Committee of the Presidium

Nanotechnology: The Path to Commercialization and Use in Everyday Life
Reyad Sawafta
President and CEO of QuarTek Corporation, a nanotechnology company in the USA

Opportunities with Soft X-Rays Synchrotron Radiation: From Nanoscience to Nanotechnology
Zahid Hussain
Senior Staff Scientist and Division Deputy for Scientific Support at the Advanced Light Source Division of E. O. Lawrence Berkeley National Laboratory (USA)

Hydrogen in Nano-Scaled Metals: Micro-Structural Aspects
Astrid Pundt
Institute of Metal Physics at the Georg-August University in Göttingen, Germany

Pt-Based Nanomaterials and their Applications in Fuel Cells
Li-Jun Wan
Institute of Chemistry, Chinese Academy of Sciences, Beijing 100080, China

THEORY

A van der Waals Type Equation of State for Confined Fluids in Nanopores

G. R. Vakili-Nezhaad

Institute for Nanoscience and Nanotechnology, University of Kashan, Kashan 87317-51167, Iran
vakili@kashanu.ac.ir

Abstract. Based on the new van der Waals equation of state for confined fluids in nanopores, an interesting thermodynamic interpretation has been made for the fragmentation phenomenon of the confined fluids in the nanopores such as carbon nanotubes. The new concept of anisotropic fugacity has been introduced in this work.

Keywords: Nanothermodynamics, Confined Fluids, Nanopores.
PACS: 64.70. -p

INTRODUCTION

In recent years the concepts of nanothermodynamics have been examined in the study of small systems [1-4]. For the study of small systems one has to improve the basic equations of the conventional (macro) thermodynamics. Although the fundamental concepts of nanothermodynamics have been introduced by T. L. Hill [5,6] in the early 1960s, there are serious challenges in defining the environmental variables in nanosystems such as temperature and pressure. Parallel to nanothermodynamics much effort has been devoted to using the concepts of non-extensive statistical mechanics and thermodynamics following the pioneering work of Tsallis in 1988 [7,8]. It has been shown that there is a close relation between nanothermodynamics and non-extensive thermodynamics via the fundamental parameters in two disciplines [9] and simulation data could determine the basic entropic parameter of non-extensive small systems by using the methods of molecular dynamics [10]. Recently it has been shown that thermodynamic equations for a confined fluid on the nanometric scale may be different from their counterparts in macrothermodynamics [11,12]. For the study of pressure-volume-temperature (P-V-T) behavior of the confined fluids in a small pore (nanopore) Zarragoicoechea and Kuz [13] have improved the van der Waals equation of state considering the tensorial nature of the pressure in the nanopores. In this work, using the modified form of the van der Waals equation of state, some thermodynamic properties have been obtained to interpret the complex phenomenon of phase change or fragmentation of the confined fluids in the nanopores.

THEORY AND DISCUSSION

Following the pioneering work of Landau and Lifshitz [14] on the assumption of pressure as a diagonal tensor, Zarragoicoechea and Kuz [13] proposed the following equations of state for a confined fluid in a long nanotube with the nanosized diameter

$$p_{xx} = p_{yy} = \frac{NkT}{V - Nb} - \frac{N^2}{V^2}\left[a - \varepsilon\sigma^3\left(3\frac{c_1}{\sqrt{A}} + 4\frac{c_2}{A}\right)\right] \tag{1}$$

CP929, *Nanotechnology and Its Applications, First Sharjah International Conference*
edited by Y. I. Salamin, N. M. Hamdan, H. Al-Awadhi, N. M. Jisrawi, and N. Tabet
© 2007 American Institute of Physics 978-0-7354-0439-7/07/$23.00

3

$$p_{zz} = \frac{NkT}{V-Nb} - \frac{N^2}{V^2}\left[a - 2\varepsilon\sigma^3\left(\frac{c_1}{\sqrt{A}} + \frac{c_2}{A}\right)\right]$$ (2)

$$c_1 = 4.6571$$ (3)

$$c_2 = -2.1185$$ (4)

where a and b are the van der Waals equation of state bulk parameters, ε and σ are energy and size parameters of the Lennard-Jones potential, and A is the cross sectional area of the nanopore. These equations recover the conventional van der Waals equation of state (in bulk) when the cross sectional area goes to infinity [13] and we have

$$p_{xx} = p_{yy} = p_{zz}$$ (5)

By using equations (1) and (2) we can use the equilibrium criterion through the calculation of fugacity of the confined fluid in different directions. We may use the general definition of the fugacity which reads [15]

$$f = p\exp\left(\int_0^p (Z-1)dp / p\right)$$ (6)

Considering the various components of the pressure tensor in different directions we can write the following direction dependent equations for the fugacity

$$f_{xx} = f_{yy} = p_{xx}\exp\left(\int_0^{p_{xx}} (Z_{xx}-1)dp_{xx} / p_{xx}\right) = p_{yy}\exp\left(\int_0^{p_{yy}} (Z_{yy}-1)dp_{yy} / p_{yy}\right)$$ (7)

and

$$f_{zz} = p_{zz}\exp\left(\int_0^{p_{zz}} (Z_{zz}-1)dp_{zz} / p_{zz}\right),$$ (8)

where Z_{xx}, Z_{yy} and Z_{zz} could be obtained by multiplication by V/RT of both sides of equations (1) and (2). After obtaining the relevant equations for the compressibility factors we can use equations (7) and (8) to derive the fugacity of a confined fluid in a nanopore for each direction. The results may be written as

$$\ln f_{xx} = \ln\frac{V-Nb}{V} - \frac{N}{kTV}\left[a - \varepsilon\sigma^3\left(3\frac{c_1}{\sqrt{A}} + 4\frac{c_2}{A}\right)\right] + Z_{xx} - 1 + \ln\frac{NkT}{V}$$ (9)

$$\ln f_{yy} = \ln\frac{V-Nb}{V} - \frac{N}{kTV}\left[a - \varepsilon\sigma^3\left(3\frac{c_1}{\sqrt{A}} + 4\frac{c_2}{A}\right)\right] + Z_{yy} - 1 + \ln\frac{NkT}{V}$$ (10)

and,

$$\ln f_{zz} = \ln\frac{V-Nb}{V} - \frac{N}{kTV}\left[a - 2\varepsilon\sigma^3\left(\frac{c_1}{\sqrt{A}} + \frac{c_2}{A}\right)\right] + Z_{zz} - 1 + \ln\frac{NkT}{V}$$ (11)

In the above equations Z_{xx}, Z_{yy} and Z_{zz} may be written as

$$Z_{xx} = Z_{yy} = \frac{V}{V-Nb} - \frac{N}{VkT}\left[a - \varepsilon\sigma^3\left(3\frac{c_1}{\sqrt{A}} + 4\frac{c_2}{A}\right)\right]$$ (12)

$$Z_{zz} = \frac{V}{V-Nb} - \frac{N}{VkT}\left[a - 2\varepsilon\sigma^3\left(\frac{c_1}{\sqrt{A}} + \frac{c_2}{A}\right)\right]$$ (13)

4

As it can be seen from equations (9)-(11) the fugacity of a confined fluid in a nanopore is an anisotropic property dependent on the geometrical direction. On the other hand, this property is applied as a criterion for fluid phase equilibria. Therefore, we can conclude that there are different phases in x-y direction from the z-direction in a long nanosized tube such as a carbon nanotube. This fact has been confirmed in the experimental observations of the fragmentation phenomenon of the confined water in long nanotubes.

REFERENCES

1. T. L. Hill, Perspective: Nanothermodynamics, Nano Lett. **1**, 111-112 (2001).
2. T. L. Hill, Extension of Nanothermodynamics to Include a One-Dimensional Surface Excess, Nano Lett. **1**, 159-160 (2001).
3. T. L. Hill, A Different Approach to Nanothermodynamics, Nano Lett. **1**, 273-275 (2001).
4. T. L. Hill, Fluctuations in Energy in Completely Open Small Systems, Nano Lett. **2**, 609-613 (2002).
5. T. L. Hill, *Thermodynamics of Small Systems*, Vol. I (Benjamin, New York, 1963).
6. T.L. Hill, *Thermodynamics of Small Systems*, Vol. II (Benjamin, New York, 1964).
7. C. Tsallis, Possible Generalization of Boltzmann-Gibbs Statistics, J. Stat. Phys. **52**, 479 (1988).
8. S. R. A. Salinas and C. Tsallis, Nonextensive Statistical Mechanics and Thermodynamics, Brazilian J. Phys. **29**, (1999).
9. G. R. Vakili-Nezhaad and G. A. Mansoori, an Application of Non-Extensive Statistical Mechanics to Nanosystems, J. Comput. Theor. Nanosci. **1**, 233-135 (2004).
10. P. Mohazzabi and G. A. Mansoori, Why Nanosystems and Macroscopic Systems Behave Differently, J. Comput. Theor. Nanosci., (in press).
11. A. G. Meyra, G. J. Zarragoicoechea and V. A. Kuz, Thermodynamic Equations for a Confined Fluid at Nanometric Scale, Fluid Phase Equilib. (in press)
12. G. J. Zarragoicoechea and V. A. Kuz, Critical Shift of a Confined Fluid in a Nanopore, Fluid Phase Equilib. **220**, 7 (2004).
13. G. J. Zarragoicoechea and V. A. Kuz, van der Waals Equation of State for a Fluid in a Nanopore, Phys. Rev. E **65**, 021110 (2002).
14. L. Landau and E. Lifshitz, *Teoria de la Elasticidad* (Reverte, Barcelona, 1969).
15. J. M. Prausnitz, R. N. Lichtenthaler, and E. G. de Azevedo, *Molecular Thermodynamics of Fluid Phase Equilibria*, Third Edition (Prentice Hall, 1999).

Study of GaAs/AlGaAs Superlattices in Structural Disorder Case

R. Djelti, S. Bentata and Z. Aziz

Département de Physique, Faculté des Sciences, Université Abdelhamid Ibn Badis, BP 227 Mostaganem 27000, Algérie
m.univ@caramail.com

Abstract. By using a transfer matrix model, we examine in this paper the extended states of a non-interacting electron system in one-dimensional Dimer/Trimer randomly placed in a superlattice. We study numerically the effects of short-range correlated disorder on the electronic and transport properties of intentionally disordered GaAs/AlxGa1-xAs superlattices. We consider layers having identical thickness where the Al concentration x takes at random two different values with the constraint that one of them appears only in pairs or triply, i.e., the random dimmer/Trimer barrier.

Keywords: superlattices, random Dimer/trimer barrier, delocalized state, transmission coefficient.
PACS: 85.30.Vw, 61.72.Ji

INTRODUCTION

Since the scaling hypothesis [1] it has been well established that all electronic states of one-dimensional (I-D) systems with random potentials are exponentially localized irrespective of the disorder. The consequence of this hypothesis is the absence of transport in an infinite disordered chain. However, recently some examples of 1-D systems with correlated disorder [2] or non-linear disordered Hamiltonians [3] have been found to exhibit a number of extended states.

Chomette et al. [4] claimed that Anderson localization [5] was responsible for the increase of the photoluminescence intensity observed in intentionally disordered SLs.

The simplest model that exhibits the suppression of localization in disordered systems is the so-called continuous random dimer model [6, 7].

In this work we consider two structures, the first one is SL where two AlxGa1-xAs barriers of different Al mole fractions are introduced at random in the SL, but one of them can only appear in pairs, this heterostructure will be referred to as dimer barrier SL, (DBSL), and the second one appear only in triply, this is trimer barrier SL, (TBSL).

FORMALISM

We study the electronic properties of the RDBSL and RTBSL in the stationary case. The one-dimensional, time-independent Schrödinger equation for an electron in a semiconductor heterostructure, with potential V (z), under the envelope function/effective mass approximations is given as

$$\left[-\frac{\hbar^2}{2}\frac{d}{dz}\frac{1}{m^*(z)}\frac{d}{dz}+V(z)\right]\Psi(z)=E\Psi(z),\tag{1}$$

Where z is the growing axis, E the incoming electron energy, $\psi(z)$ the wave function in the growing direction and m* the effective mass of each monolayer. We solve Eq. (1) by using the transfer matrix formalism [8] and the Bastard continuity conditions [9], for an incident electron coming from the left one has the relation between the reflected and transmitted amplitude r and τ, respectively

$$\begin{pmatrix}1\\r\end{pmatrix}=M(0,L)\begin{pmatrix}\tau\\0\end{pmatrix}.\tag{2}$$

A simple algebra yields the transfer matrix M (0,L) as

$$M(0,L)=-\frac{m_w^*}{2ik}\begin{pmatrix}-\dfrac{ik}{m_w^*}&-1\\[2mm]-\dfrac{ik}{m_w^*}&1\end{pmatrix}S(0,L)\begin{pmatrix}1&1\\[2mm]\dfrac{ik}{m_w^*}&-\dfrac{ik}{m_w^*}\end{pmatrix}.\tag{3}$$

Here the diffusion matrix S (0, L) can be formulated in terms of the elementary diffusion matrices Gj(l) associated to each region j of the potential having a width l as the product

$$S(0,L)=G_j(l)=\begin{pmatrix}S_{11}&S_{12}\\S_{21}&S_{22}\end{pmatrix}.\tag{4}$$

The transmission coefficient is then given by

$$T=\frac{4}{(S_{11}+S_{22})^2+\left(\dfrac{k}{m_w}S_{12}-\dfrac{m_w}{k}S_{21}\right)^2}\tag{5}$$

This expression measures the electron interaction with the structure through the elements of the diffusion matrix S(0,L) and the wave vector defined by $k^2=\dfrac{2m_w}{\hbar^2}E$.

We consider SL constituted by two semiconductor materials the well width d_w is different than the barrier thickness d_b in the whole sample which in turns preserves the periodicity of the lattice along the growing axis. $d=d_w+d_b$ is the period of the unit supercell. The system consists of N rectangular barriers (in this paper N = 200) .In this model of disordered SL, we consider that the height of the barriers takes at random only two values, namely V and $\overline{V}$. These two potentials height are proportional to the two possible values of the Al fraction x in the AlxGa1-xAs barriers.

The sequence of potential is short-range correlated since the $\overline{V}$ appears forming triply randomly placed in the structure, i.e., $V\overline{VV}\ VV\overline{VV}\ V\overline{VV}\ V\ldots\ldots\ldots$.

RESULTS AND DISCUSSION

This section concerns the statistical description of the electronic transport process in the RDBSL and RTBSL mesoscopic devices, by means of numerical calculations of its transmission coefficient with the corresponding Lyapunov Exponant.

Physical parameter values, such as dw = 20 ° A, db = 26 ° A, V = 220 meV and $\overline{V}$ = 300 meV, are chosen, to obtain allowed minibands lying below the barriers. The corresponding effective masses are taken to be m_w = 0.067m_0, m_b = 0.096m_0 and m_b = 0.089m_0 for respectively the quantum well, host barrier and dimer/Trimer barrier layers where m_0 is the free electron mass. For convenience, the bottom of GaAs wells has been chosen as the energy reference.

For the above parameters, transmission coefficient versus electron incident energy $\tau(E)$ is plotted. Fig. 1 shows the position of the lower and upper band edges of the minibands corresponding to the two ordered superlattices with the two barrier heights V and $\overline{V}$. One can observe the existence of one miniband under the well, ranging from 114 up to 256 meV for V and from 146 up to 259 for $\overline{V}$.

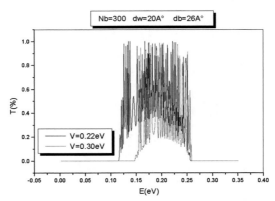

FIGURE 1. Transmission coefficient versus incident electron energy for two structures ordonned.

The resonant energies lie inside the common region between the allowed minibands of V and V. This provides the existence of different types of eigenstates: those having low resistances near resonant energies and those, with considerable resistances, very far from the quasi perfect energy transparencies (i.e. near band tails and inside the valley mini gap). This result is in perfect accord with that obtained by Bentata et al. [10].

In Fig. 2 it appears that for c = 0.15 and c = 0.25, the position of such resonant energies does not depend on the degree of disorder. One can conclude that their positions obey directly to dimer/trimer and host unit cell technological characteristics (i.e., aluminium molar fraction x in the AlxGa1-xAs layers).

We have found three resonant peaks in RTBSL, two of them are due to the trimer tunnelling state originated for the basic cells of three singular barriers (figure.3), unexpectedly, the TRBSL support another type of resonance its origin being completely different, this one is due to the commutation of the two potential V and $\overline{V}$.

We represent in figure 4 the Lyanopov exponent versus the energy for the TRBSLand DRBSL with two different disorder concentration C=0.15 and c=0.25.

We observe the existence of different types of eigenstates: those having a very low lyapunov exponent close to the resonant energy and those with high slope in other region.

The particular behaviour of lyapunov exponent suggests the existence of a cross-over between separating the two phases: strongly localized states in the band tails from weakly or extended localized near to the resonant energy.

We noted that the delocalized states due to commutation are not clean only with the dimer but also for the trimer [11].

8

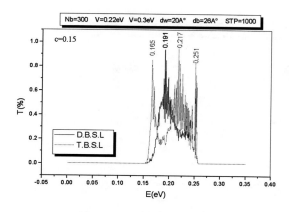

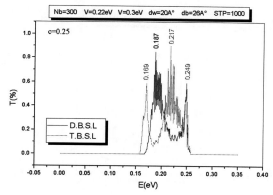

FIGURE 2. Transmission Coefficient versus electron energy for two disorder concentration.

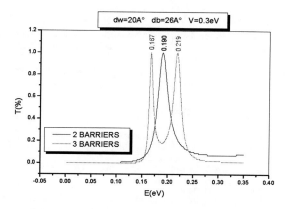

FIGURE. 3. Transmission Coefficient versus electron energy for elementary cells.

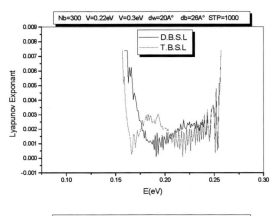

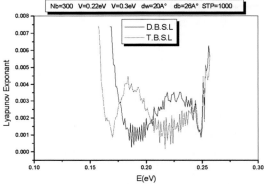

FIGURE.4: Lyapunov Exponent versus electron energy for D.B.S.L and T.B.S.L disorder concentration: a) c=0.15, and b) c=0.25.

CONCLUSION

We presented two fundamental ingredients to thwart the destructive influence of the disorder: Same period along the increasing axis maintains of superlattice and short carried disorder correlation. These two devices create the conditions supporting electronic transport by resonant tunnel effect.

In summary we observed that the introduction of a disorder correlated into the superlattices prevent the localization and causes delocalization wide states. The trimer height barriers structure presents wider delocalization than the dimer.

REFERENCES

1. E. Abrahams, P. W. Anderson, D. C. Licciaardello and T. V. Ramakrishnan, Phys. Rev. **42**, 673 (1979).
2. D. Dunlap, H-L. Wu and P. Phillips, Phys. Rev. Lett. **65**, 88 (1990).
3. M. I. Molina and G.P. Tsironis, Phys. Rev. Lett. **73**, 464 (1994).
4. A. Chomette, B. Deveaud, A. Regreny, and G. Bastard, Phys. Rev. Lett. **57**, 1464 (1986).
5. P. W. Anderson, Phys. Rev. **109**, 1492 (1958).
6. J. C. Flores, J. Phys.: Condens. Matter **1**, 8471 (1989).
7. D. H. Dunlap, H.-L. Wu, and P. Phillips, Phys. Rev. Lett. **65**, 88 (1990).

8. J. S. Walker and J. Gathright, Amer. J. Phys. **62**, 408 (1994).
9. G. Bastard, Phys. Rev. B**24**, 5693 (1981).
10. S. Bentata, *Superlattices and Microstructures* **37**, 297 (2005).
11. R. Djelti, S.Bentata and Z. Aziz, FIZIKA A **15**, 219 (2006).

Distance Matrix and Wiener Index of Polyhex Nanotubes

A. R. Ashrafi and S. Yousefi

Institute of Nanoscience and Nanotechnology, University of Kashan, Kashan 87317-51167, Iran
Center for Space Studies, Malek-Ashtar University of Technology, Tehran, Iran
alir.ashrafi@gmail.com

Abstract. A topological index is a numeric quantity that is mathematically derived in a direct and unambiguous manner from the structural graph of a molecule. It has been found that many properties of a chemical compound are closely related to some topological indices of its molecular graph. Among topological indices, the Wiener index is probably the most important one. This index was introduced by the chemist H. Wiener (1947) about 60 years ago to demonstrate correlations between physicochemical properties of organic compounds and the topological structure of their molecular graphs. He defined his index as the sum of distances between any two carbon atoms in the molecules, in terms of carbon-carbon bonds. The aim of this paper is to prepare some algorithms for computing distance matrix and Wiener index of zig-zag and armchair polyhex nanotubes.

Keywords: Distance matrix, Wiener index, zig-zag polyhex nanotube, armchair polyhex nanotube.
PACS: 81.07.De; 81.07.Nb; 81.16.Rf.

INTRODUCTION

Carbon nanotubes form an interesting class of carbon nanomaterials. These can be imagined as rolled sheets of graphite about different axes. These are three types of nanotubes: armchair, chiral and zigzag structures. Further nanotubes can be categorized as single-walled and multi-walled nanotubes and it is very difficult to produce the former.

Graph theory has found considerable use in chemistry, particularly in modeling chemical structure. Graph theory has provided the chemist with a variety of very useful tools, namely, the topological index. A topological index is a numeric quantity that is mathematically derived in a direct and unambiguous manner from the structural graph of a molecule. It has been found that many properties of a chemical compound are closely related to some topological indices of its molecular graph.

Among topological indices, the Wiener index is probably the most important one, see [1]. This index was introduced by the chemist H. Wiener about 60 years ago to demonstrate correlations between physicochemical properties of organic compounds and the topological structure of their molecular graphs. He defined his index as the sum of distances between any two carbon atoms in the molecules, in terms of carbon-carbon bonds. Next Hosoya named such graph invariants, topological index, [2].

The fact that there are good correlations between and a variety of physico-chemical properties of chemical compounds containing boiling point, heat of evaporation, heat of formation, chromatographic retention times, surface tension, vapor pressure and partition coefficients could be rationalized by the assumption that Wiener index is roughly proportional to the van der Waals surface area of the respective molecule.

Diudea was the first chemist which considered the problem of computing topological indices of nanostructures. He and his co-authors computed the Wiener index of an armchair and zig-zag polyhex nanotubes, [3,4]. The present authors computed the Wiener index of a polyhex and TUC4C8(R/S) nanotori [5,6]. In this paper, a matrix method is described,

CP929, *Nanotechnology and Its Applications, First Sharjah International Conference*
edited by Y. I. Salamin, N. M. Hamdan, H. Al-Awadhi, N. M. Jisrawi, and N. Tabet
© 2007 American Institute of Physics 978-0-7354-0439-7/07/$23.00

by means of which it is possible to calculate some topological indices of nanotubes. Using this method, the Wiener index of zig-zag and armchair polyhex nanotubes are computed, see Figures 1 and 2.

Throughout this paper, our notation is standard and taken mainly from [7] and standard books of graph theory. Computations were carried out with the aid of TOPOCLUJ [8]. We encourage reader to consult [9-12] and references therein for further study and background material on this topic.

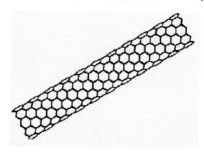

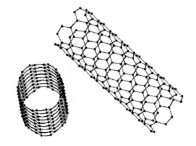

FIGURE 1. zig-zag polyhex nanotube.   FIGURE 2. Armchair polyhex nanotube.

MAIN RESULTS AND DISCUSSION

In this section two algorithms for computing distance matrix and Wiener index of armchair and zig-zag polyhex nanotubes are prepared, Figures 1 and 2. We assume that T and S denote armchair and zig-zag polyhex nanotubes with exactly m rows and n columns, respectively.

Distance Matrix and Wiener Index of an Armchair Polyhex Nanotube

Diudea, Stefu, Pârv and John [4] computed the Wiener index of an armchair polyhex nanotube $T = TUVC_6[m,n]$, for the first time. Here n is the number of vertical zig-zags and m is the number of rows, see Figure 3. It is obvious that n is even and $|V(T)| = mn$. In this section, distance matrix and Wiener index of these nanotubes are computed.

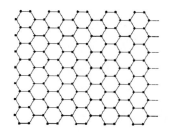

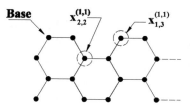

FIGURE 3. The 2–dimensional fragments of an armchair polyhex nanotube.

We first choose a base vertex b from the 2–dimensional lattice of $T = TUVC_6[m,n]$, Figure 3 and assume that $x_{i,j}^{(1,1)}$ is distance between (1,1) and (i,j). This defines a matrix $X_{m\times n}^{(1,1)}=[x_{i,j}^{(1,1)}]$, where $x_{1,1}^{(1,1)}=0$, $x_{1,2}^{(1,1)}=x_{2,1}^{(1,1)}=1$. It is clear that by choosing different base vertices, we find different distance matrices of T. Suppose $s_k^{(p,q)}$ is sum of kth row of $X_{m\times n}^{(p,q)}$, where (p,q) is the base vertex. Then $s_k^{(p,1)}=s_k^{(p,q)}$, $1\le k\le m$, $1\le p\le m$ and $1\le q\le n$. On the other hand, by choosing a fixed column, we have $s_k^{(i,j)}=\begin{cases} s_{i-k+1}^{(1,1)} & 1\le k\le i\le m,\ 1\le j\le n \\ s_{k-i+1}^{(1,1)} & 1\le i\le k\le m,\ 1\le j\le n \end{cases}$.

We now define two matrices $A_{(n/2+1)\times n}=[a_{ij}]$ and $B_{(n/2+1)\times n}=[b_{ij}]$ by $a_{1,1}=0$, $a_{2,1}=1$. If $2|j$ then $a_{1,j}=\begin{cases} a_{1,j-1}+1 & j\le (n/2)+1 \\ a_{1,j-1}-1 & j>(n/2)+1 \end{cases}$ and $a_{2,j}=\begin{cases} a_{1,j}+1 & j\le (n/2)+1 \\ a_{1,j}-1 & j>(n/2)+1 \end{cases}$. If $2\nmid j$ then $a_{2,j}=\begin{cases} a_{2,j-1}+1 & j\le (n/2)+1 \\ a_{2,j-1}-1 & j>(n/2)+1 \end{cases}$ and $a_{1,j}=\begin{cases} a_{2,j}+1 & j\le (n/2)+1 \\ a_{2,j}-1 & j>(n/2)+1 \end{cases}$. Other entries of this matrix is obtained from the first two rows by $a_{i,j}=a_{1,j}$, i is odd and $a_{i,j}=a_{2,j}$, i is even. We also define $b_{n/2+1,j}=\begin{cases} n/2+j-1 & j\le n/2+1 \\ 3n/2-j+1 & j>n/2+1 \end{cases}$ and other entries of B is defined by equation $b_{i,j}=b_{i+1,j}-1$, $i < n/2 + 1$. Therefore, $x_{i,j}^{(1,1)}=\begin{cases} c_{i,j} & i\le (n/2)+1 \\ x_{i-1,j}^{(1,1)}+1 & i>(n/2)+1 \end{cases}$, where $c_{i,j} = Max\{a_{i,j}, b_{i,j}\}$. Since the Wiener index is equal to one half of the sum of entries of distance matrix, so we have the following theorem:

Theorem 1. Suppose T is an armchair polyhex nanotube with exactly m rows and n columns, Figure 2, and $W = W(T)$. Then,

$$W=\begin{cases} \dfrac{m^2n}{12}\left(3n^2+m^2-4\right)+\dfrac{n}{8}(-1)^{(n/2)}\left[1-(-1)^m\right] & m\le \dfrac{n}{2}+1 \\ \dfrac{mn^2}{24}\left(n^2+4m^2+3mn-8\right)-\dfrac{n^3}{192}\left(n^2-16\right)+\dfrac{n}{8}\left[(-1)^{(n/2)}-1\right] & m>\dfrac{n}{2}+1 \end{cases} \tag{1}$$

Distance Matrix and Wiener Index of Zig–Zag Polyhex Nanotubes

John and Diudea [3] computed the Wiener index of a zig-zag polyhex nanotube $T = TUHC_6[m,n]$, where n is the number of vertical zig-zags and m is the number of rows, see Figures 1 and 4. It is obvious that n is even and $|V(T)| = mn$. We first choose a base vertex b from the 2–dimensional lattice of T and assume that xij is the $(i,j)^{th}$ vertex of T, Figure 4. Define $D_{m\times n}^{(1,1)}=[d_{i,j}^{(1,1)}]$, where $d_{i,j}^{(1,1)}$ is distance between (1,1) and (i,j), i = 1,2...,m and j = 1,2,...,n. By Figure 2, there are two separates cases for the $(1,1)^{th}$ vertex. For example in the case (a) of Figure 4, $d_{1,1}^{(1,1)}=0$, $d_{1,2}^{(1,1)}=d_{2,1}^{(1,1)}=1$ and in case (b), $d_{1,1}^{(1,1)}=0$, $d_{1,2}^{(1,1)}=1$, $d_{2,1}^{(1,1)}=3$. In general, we assume that $D_{m\times n}^{(p,q)}$ is distance matrix of T related to the vertex (p,q) and $s_i^{(p,q)}$ is the sum of ith row of $D_{m\times n}^{(p,q)}$. Then there are two distance matrix related to (p,q) such that $s_i^{(p,2k-1)}=s_i^{(p,1)}$; $s_i^{(p,2k)}=s_i^{(p,2)}$; $1\le k\le n/2$, $1\le i\le m$, $1\le p\le m$.

By Figure 4 and previous notations, if b varies on a column of T then the sum of entries in the row containing base vertex is equal to the sum of entries in the first row of $D_{m\times n}^{(1,1)}$. On the other hand, one can compute the sum of entries in other rows by distance from the position of base vertex. Therefore, if $2 | (i + j)$ then $s_k^{(i,j)}=\begin{cases} s_{i-k+1}^{(1,1)} & 1\le k\le i\le m,\ 1\le j\le n \\ s_{k-i+1}^{(1,2)} & 1\le i\le k\le m,\ 1\le j\le n \end{cases}$ and if $2\nmid(i+j)$ then $s_k^{(i,j)}=\begin{cases} s_{i-k+1}^{(1,2)} & 1\le k\le i\le m,\ 1\le j\le n \\ s_{k-i+1}^{(1,1)} & 1\le i\le k\le m,\ 1\le j\le n \end{cases}$.

We now describe our algorithm to compute distance matrix of a zig-zag polyhex nanotube. To do this, we define matrices $A^{(a)}_{m\times(n/2+1)}=[a_{ij}]$, $B_{m\times(n/2+1)}=[b_{ij}]$ and $A^{(b)}_{m\times(n/2+1)}=[c_{ij}]$ as in Table 1.

For computing distance matrix of this nanotube we must compute matrices $D^{(a)}_{m\times n}=[d^a_{i,j}]$ and $D^{(b)}_{m\times n}=[d^b_{i,j}]$. But by our calculations, we can see that the following equations are satisfied

$$d^a_{i,j}=\begin{cases} \text{Max}\{a_{i,j},b_{i,j}\} & 1\le j\le \dfrac{n}{2} \\ d_{i,n-j+2} & j>\dfrac{n}{2}+1 \end{cases}, \quad d^b_{i,j}=\begin{cases} \text{Max}\{a_{i,j},c_{i,j}\} & 1\le j\le \dfrac{n}{2} \\ d_{i,n-j+2} & j>\dfrac{n}{2}+1 \end{cases}. \tag{2}$$

This completes calculation of distance matrix. By our calculations and this fact that the Wiener index is equal to one half of the sum of entries of distance matrix, we have:

Theorem 2. Suppose S is a zig-zag polyhex nanotube with exactly m rows and n columns, Figure 4. Then,

$$W_{m\times n}=\begin{cases} \dfrac{mn^2}{24}\left(4m^2+3mn-4\right)+\dfrac{m^2n}{12}\left(m^2-1\right) & m\le \dfrac{n}{2}+1 \\ \dfrac{mn^2}{24}\left(8m^2+n^2-6\right)-\dfrac{n^3}{192}\left(n^2-4\right) & m>\dfrac{n}{2}+1 \end{cases}. \tag{3}$$

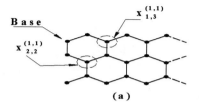

(a)

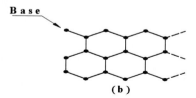

(b)

FIGURE 4. Two basically different cases for the Vertex b.

TABLE 1. Definition of three matrices $A^{(a)}_{m\times(n/2+1)}$, $B_{m\times(n/2+1)}$ and $A^{(b)}_{m\times(n/2+1)}$.

Matrix $A^{(a)}_{m\times(n/2+1)}$	$a_{1,1}=0$; $a_{1,2}=1$; $a_{i,j}=\begin{cases} a_{i,1} & 2\nmid j \\ a_{i,2} & 2\mid j \end{cases}$; $a_{i,1}=a_{i-1,1}+1$, $a_{i,2}=a_{i,1}+1$, $2\mid i$; $a_{i,2}=a_{i-1,2}+1$, $a_{i,1}=a_{i,2}+1$, $2\nmid i$
Matrix $A^{(b)}_{m\times(n/2+1)}$	$c_{1,1}=0$; $c_{1,2}=1$; $c_{i,j}=\begin{cases} c_{i,1} & 2\nmid j \\ c_{i,2} & 2\mid j \end{cases}$; $c_{i,2}=c_{i-1,2}+1$, $c_{i,1}=c_{i,2}+1$, $2\mid i$; $c_{i,1}=c_{i-1,1}+1$, $c_{i,2}=c_{i,1}+1$, $2\nmid i$
Matrix $B_{m\times(n/2+1)}$	$b_{i,1}=i-1$, $b_{i,j}=b_{i,j-1}+1$

ACKNOWLEDGMENTS

The authors would like to thank Professor Mircea V. Diudea for giving us a copy of his software TopoCluj during Math/Chem/Comp 2006 conference at Dubrovnik, Croatia.

REFERENCES

1. H. Wiener, J. Am. Chem. Soc. **69**, 17-20 (1947).
2. H. Hosoya, Bull. Chem. Soc. Japan, **44**, 2332-2339 (1971).
3. P. E. John and M. V. Diudea, Croat. Chem. Acta 77, 127-132 (2004).
4. M. V. Diudea, M. Stefu, B. Pârv and P. E. John, Croat. Chem. Acta 77, 111-115 (2004).
5. S. Yousefi and A. R. Ashrafi, MATCH Commun. Math. Comput. Chem. **56**, 169-178 (2006).
6. A. R. Ashrafi and S. Yousefi, MATCH Commun. Math. Comput. Chem. **57**, 403-410 (2007).
7. N. Trinajstic, Chemical Graph Theory, Boca Raton, FL: CRC Press, 1983.
8. M. V. Diudea, O. Ursu and C. L. Nagy, TOPOCLUJ 2.0-Calculations in MOLECULAR TOPOLOGY, B-B University, 2002.
9. A. A. Dobrynin, R. Entringer and I. Gutman, Acta Appl. Math., **66**, 211-249 (2001).
9. A. A. Dobrynin, I. Gutman, S. Klavžar and P. Zigert., *Acta Appl. Math.*, **72**, 247-294 (2002).
10. A. R. Ashrafi and S. Yousefi, Nanoscale Research Letters **2**, 202 (2007).
11. S. Yousefi and A. R. Ashrafi, submitted for publication.

MODELING AND COMPUTATIONAL

Calculation of the Heat Capacities of Solid Phase Buckminsterfullerene Using Semi-Empirical Methods

G. R. Vakili-Nezhaad, M. Hamedanian and M. Ghomashi

Institute for Nanoscience & Nanotechnology, University of Kashan, Kashan 87317-51167, Iran
vakili@kashanu.ac.ir

Abstract. Calculation of vibrational spectra for the large molecule of fullerene, C_{60}, has been carried out by efficient quantum mechanical methods. In this work the ab initio calculation methods which have been chosen were GAUSSIAN 98 program with the STO-3G basis set and HYPER 7 program with PM3 and AM1. The heat capacity at constant volume has been calculated based on the separation of the vibrational spectrum into group and molecular vibrations. Also the heat capacity at constant pressure was obtained by the modified equation of Nernst-Lindemann.

Keywords: fullerene, specific heat, computational methods.
PACS: 61.48.$^{+}$c

INTRODUCTION

Buckminsterfullerene (or fullerene), C_{60}, is a new allotrope of carbon (after graphite and diamond), which was discovered in 1985 by Kroto and collaborators[1] by the method of laser evaporation of graphite[2]. Since the time of discovery of fullerenes, a great deal of investigations has gone into these interesting and unique nanostructures and their derivatives due to their application in nanotechnology[3]. Characterization of fullerenes requires complete vibrational assignment compatible with the experimental data[4].

There are several methods to calculate the frequencies of C_{60} which can be divided into two main groups, namely semi-empirical and ab initio methods[5]. The semi-empirical model calculations of the fundamental frequencies are rather straightforward, but the results provide only a qualitative understanding of the spectrum[6-8]. With the rapid progress in computer speed and memory capacity the molecule became a testing ground for cutting edge ab initio work[9]. In this work for determining the frequencies of C_{60} the calculations were carried out first using the semi-empirical Parametric Method 3 (PM3)[10] and Austin Method 1 (AM1)[11] by HYPER7, then by using GAUSSIN 98 program with the basis set of STO-3G. Based on the calculated frequencies, the heat capacities have been obtained.

MODELING AND CALCULATION

All quantum mechanical calculations in this work have been performed with ab initio GAUSSIAN 98 and HYPER 7 programs. Calculations of dipole moments (DM) utilizing GAUSSIN 98 were performed with the STO-3G basis set. Calculations of molecular DM utilizing HYPER 7 were performed with PM3 and AM1. It is an important point to say that the structure of C_{60} has been optimized by the method of RHF with basis sets of STO-3G, 4-31G, 6-31G, D95 and 6-31G**, but due to some computational limits only the basis set of STO-3G could calculate the vibrational frequencies. By using the calculated vibrational frequencies above, the heat capacity at constant volume (C_v) is obtained by the following equation

CP929, Nanotechnology and Its Applications, First Sharjah International Conference
edited by Y. I. Salamin, N. M. Hamdan, H. Al-Awadhi, N. M. Jisrawi, and N. Tabet
© 2007 American Institute of Physics 978-0-7354-0439-7/07/$23.00

$$C_V = R \times \sum_{i=1}^{46} g_i (\theta_i / T)^2 \frac{\exp(\theta_i / T)}{(\exp(\theta_i / T) - 1)^2}, \qquad (1)$$

where R is the gas constant, 8.31454 (J/mole K), T is the temperature in K, and $\theta_i = h\nu_i / K$ is the characteristic frequency expressed in K (a frequency ν_i given, as customary in spectroscopy, in wave numbers, cm^{-1}, must be multiplied by 1.4388 cm K to yield the frequency in K). The conversion of heat capacity at constant volume to that at constant pressure (C_P) is done by the modified Nernst-Lindemann equation[12]

$$C_P - C_V = \frac{3RA^0 C_p T}{T_m^0}, \qquad (2)$$

where A° is an experimental coefficient with the value of 3.9×10^{-3} kmol/J, and T^0_m represents the equilibrium melting temperature which is estimated to be 1000 K[13].

A rapid jump has been observed in the heat capacity curve due to the phase transition of crystalline to the plastic crystal solid phases (see Fig. 1). The solid phase transition has occurred in the temperature range of 260-270 K which is totally in agreement with the experimental phase transition temperature (260.7 K).

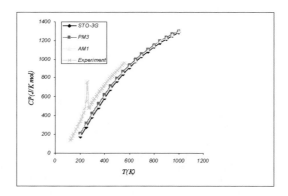

FIGURE 1. Heat capacity (C_P) vs T by using of Calculated Vibrational Frequancies

RESULTS AND DISCUSSION

In this work, using GAUSSIAN 98 program with the STO-3 G basis set and HYPER 7 program with PM3 and AM1, the heat capacity of C_{60} at constant volume (C_V) has been calculated. Also, using the modified Nernst-Lindemann equation, the heat capacity of C_{60} at constant pressure (C_P) has been calculated. Also such differences may be found from the deviation between calculated and measured heat capacities at constant volume and heat capacities at constant pressure. Moreover, the measured value of C_{60} heat capacity (C_P) agrees reasonably well with the calculated heat capacities based on the various vibrational frequencies. Since three sets of normal mode calculations of group vibrations were reported independently, in this work these have been combined with the molecular vibrations to calculate the heat capacity. In a short interval of temperature around 260 K some major differences between the calculated and experimental results have been observed and this rapid jump in the C_P curve goes back to the phase transition of crystalline to the plastic crystal solid phases.

REFERENCES

1. H. W. Kroto, J. R. Heath, S. C. O. Brien, R. F. Curl, and R. E. Smalley, Nature B18, 165 (1985).
2. G. A. Mansoori, UN-APCTT Tech Monnitor, Special Issue, pp. 53-59, Sep.-Oct. (2002).
3. W. Krätshmer, L. D. Lamb, K. Fostiropoulos, and D. R. Huffman, Nature 347, 354 (1990).
4. Long, V. C. Lang, J. I. Musfeldt, K. Kamaras, A. Schilder, and W. Schutz, Phys. Rev. B 58, 14338 (1998).
5. Ira-N. Levine, Quantum Chemistry (1990).
6. F. Negri, G. Orlandi and F. Zerbetto, J. Am. Chem. Soc. 1136, 7 (1991).
7. R. A. Jishi, R. M. Mirie and M. S. Dresselhaus, Phys. Rev. B 45, 13685-9 (1992).
8. A. A. Quong, M. R. Pederson, and J. L. Feldman, Solid State Commun. 87, 535-8 (1993).
9. C. H. Choi, M. L. Kertesz, J. Phys. Chem. A 104, 102-12 (2000).
10. Y. Shinohara, R. Saito, T. Kimura, G. Dresselhaus and M. S. Dresselhaus, Chem. Phys. Lett. 227, 365, (1994).
11. Z. Slanina, M. Rudzinski, M. Togasi and E. Osawa, J. Mol. Struct. Theochem 202, 169 (1989).
12. Yimin Jin, Jinlong Cheg, Manika Varma-Nair, Guangh Liang, Yigang fu, Bernhard Wunderlich, J. Phys. Chem. 96, 5151-5156 (1992).
13. K. A. Wang, A. Rao, M. Eklund, P. C. Dresselhaus, M. S., and G. Dresselhaus, Phys. Rev. B 48, 11375 (1993).

Periodic Nanostructured Pins and the Anomaly in the Critical Depinning Force

I. M. Obaidat[1], U. Al Khawaja[1], M. Benkraouda[1], F. Hamed[1] and N. Salmeen[2]

[1] Department of Physics, United Arab Emirates University, Al-Ain 17551, United Arab Emirates
[2] Al Nasha'a Al Saleh Private School, Al-Ain, United Arab Emirates
lObaidat@uaeu.ac.ae

Abstract. For the first time, molecular dynamic simulations based on dense nanostructured periodic arrays of pinning sites were carried out. The variables in the simulations were the vortex density, the temperature, the pinning strength, the size of pinning sites. An interesting *dip* was found to occur in the critical current density but only at zero temperature and for specific values of pinning strength. The properties of the *dip* were found to depend strongly on the initial positions of the vortices with respect to the positions of the pinning sites. The occurrence of the dip is attributed to the formation of one dimensional linear channels of moving vortices in the direction of the applied current. The pinning strength was also found to have an important role in determining the relative depth of the *dip*.

Keywords: Periodic Nanostructures, Critical current density, Dip effect, Simulations.
PACS: 74.25.Qt, 74.25.Sv.

INTRODUCTION

It is crucial that the critical current density, J_c is made as high as possible for any practical applications of high temperature superconductors (HTSCs). Increasing the applied magnetic field or the temperature was found to suppress the pinning forces which in turn lowered J_c [1]. However a sharp enhancement of J_c with increasing the applied magnetic field was reported for several superconductors [2-5]. Such anomalous behavior has been termed the peak effect (or the fish-tail). The competition between the repulsive vortex-vortex interactions and the attractive vortex-pin interactions have resulted in this striking effect. There has been a general consensus that the occurrence, location, strength, and width of the peak depend on the size [6] and strength [7] of the pinning sites. Numerical studies of the effect of the strength of the pinning sites on the peak effect in superconductors with random distribution of pinning sites were conducted [8], it was found that the peak effect is more pronounced in the weak pinning limit and progressively diminishes with increasing pinning strength.

The advancement of electron lithography enabled the creation of well-defined nanostructured periodic arrays of pinning arrays. It is possible to construct samples with well-defined periodic pinning structures in which the microscopic pinning parameters, such as size, depth, periodicity, and density, can be carefully controlled [9]. Periodic pinning arrays are also of technological importance since the arrays can produce higher critical currents than an equal number of randomly placed pins [10, 11]. Simulations on systems of square arrays of pinning sites [12] showed that the occurrence, location and height of the peak depend on the size of pinning sites. While numerical studies [13] on systems with small number of pinning sites showed that the occurrence of the peak and its height depend on the distribution of the pinning sites. In the case of square array of pinning sites, the occurrence of the peak was found at most matching and fractional matching of the first applied magnetic field. In both of the above studies [12, 13], the

CP929, *Nanotechnology and Its Applications, First Sharjah International Conference*
edited by Y. I. Salamin, N. M. Hamdan, H. Al-Awadhi, N. M. Jisrawi, and N. Tabet
© 2007 American Institute of Physics 978-0-7354-0439-7/07/$23.00

peak in the critical current occurred at those matching fields where the interstitial vortices can form a highly ordered lattice.

In the present paper we report on the on the appearance of a "dip" effect in the critical current density in a very dense nanostructured periodic pinning sites of pinning density $n_p = 0.74\lambda^{-2}$. This high density of nanostructured pinning centers can only be achieved using X-ray interference lithography technique [14], where nanostructures with periods as small as *40 nm* have been produced. Similar to the properties of the peak effect, the temperature, pinning strength and vortex initial positions were found to influence the occurrence and properties of the dip effect.

SIMULATIONS

We consider a *2D* transverse slice (in the *xy*-plane) of an infinite *3D* slab containing rigid vortices and columnar defects, all parallel to both the sample edge and the applied field $\mathbf{H} = H\hat{z}$. These vortices attain a uniform density n_v, allowing us to define the external field $H = n_v\phi_o$, where $\phi_o = hc/2e$. This model is most relevant to superconductors with periodic arrays of columnar defects or thin-film superconductors where the vortices can be approximated by *2D* objects. We model the vortex-vortex force by a modified Bessel function of the first kind, $K_1(r\backslash\lambda)$, where λ is the penetration depth [2, 3]. For the vortex-pin interaction, we assume the pinning potential well to be parabolic [2, 3]. The pinning range (i.e., the radius of the parabolic well) is r_p. For computational efficiency the vortex-vortex interaction can be safely cut off at 6λ since the Bessel function decays exponentially for r grater than λ. We use finite temperature overdamped molecular dynamics simulations in two dimensions. The overdamped equation of motion for each vortex is given by

$$\mathbf{f}_i^{tot} = \mathbf{f}_i^{vv} + \mathbf{f}_i^{vp} + \mathbf{f}_i^T + \mathbf{f}_d = \eta\mathbf{v}_i, \tag{1}$$

where $\mathbf{f}_i^{tot}$ is the total force on vortex i, $\mathbf{f}_i^{vv}$ is the vortex-vortex force, $\mathbf{f}_i^{vp}$ is the vortex-pin force, f_d is the driving force in the *x*-direction corresponding to the Lorentz force, and the thermal fluctuations are accounted for by a stochastic term that has the properties $\langle f_i^t \rangle = 0$ and $\langle f_i^T(t) f_j^T(t') \rangle = 2\eta k_B T\delta(t-t')\delta_{ij}$, where f_i^T is the thermal force given by $f_i^T = Af_o$, and A is the number we tune to vary T. In this manner the temperature is given by $T = 1/(2\eta k_B)(Af_o)^2\Delta t$, where Δt is the time step used in the numerical simulation [2, 3]. The total force on each vortex, due to all other vortices and pinning sites can be expressed as follows

$$f_i = \sum_{j=1}^{N_v} f_o K_1\left(\frac{r_i - r_j}{\lambda}\right)\hat{r}_{ij} + \sum_{k=1}^{N_p} \frac{f_p}{r_p}\left|r_i - r_k^p\right|\Theta\left(\frac{r_p - \left|r_i - r_k^p\right|}{\lambda}\right)\hat{r}_{ik}. \tag{2}$$

Here Θ is the Heaviside step function, r_i is the location of the ith vortex, r_k^p is the location of the kth pinning site, $\hat{r}_{ij} = (r_i - r_j)/|r_i - r_j|$, $\hat{r}_{ik} = (r_i - r_k^p)/|r_i - r_k^p|$, f_p is the pinning strength, N_v is the number of vortices, and N_p is the number of pinning sites. We measure all forces in units of $f_o = \phi_o^2/8\pi^2\lambda^3$, fields in units of ϕ_o/λ^2, lengths in units of λ, temperature in units of $\lambda f_o/k_B$, and the velocity in units of f_o/η. We take $f_o = k_B = \eta = 1$.

Our system has a size of $36\lambda \times 36\lambda$. The pinning sites are distributed over this area in a square array with a density $n_p = 0.74\lambda^{-2}$.

Initially, we place the vortices in a perfect lattice subject to a uniform driving force f_d in the positive x-direction (which would correspond to a Lorentz force due to an applied current). For each value of the driving force F_d, the average velocity $\overline{v}_x$ of all vortices is calculated after the steady-state is reached

$$\overline{v}_x = \frac{1}{N_v} \sum_{i=1}^{N_v} \mathbf{v}_i \cdot \hat{x}, \qquad (3)$$

The average velocity $\overline{v}_x$ versus the force f_d curve corresponds experimentally to a voltage-current, $V(I)$, curve. The critical depinning force F_d^c corresponds to the critical current density marking a transition from the pinned to the moving vortex phase. We used the Euler method to solve the equations of motion. The time step used is $\Delta t = 0.02$. we found that the maximum time needed for the vortices to reach a steady state is 2×10^4 for all of our calculations.

RESULTS

Figure 1 represents a plot of the calculated average velocity of the vortices versus the driving Lorentz force at different temperatures. At zero temperature, the average velocity remains zero up to the value of the critical depinning force F_d^c. Above F_d^c, the vortices are depinned and flow. Similar results were obtained when the size of the array, the density of pinning sites, the strength of pinning and the size of pinning sites are varied. We have observed an anomaly in the critical depinning force, which we termed the "dip effect" in analogy to the well known "peak effect". This dip effect was found to occur only at zero temperature. Figure 2 is a plot of the critical depinning force F_d^c as function of B/B_ϕ for several values of the pinning force strength f_p, where B is the magnetic flux density of the vortices and B_ϕ is a flux density corresponding to the pinning sites. It is seen that F_d^c decreases as the density of vortices is increased; however, a clear dip in this general behavior of F_d^c occurs.

We have found that this dip is mostly pronounced for $f_p = 2.0$, $f_p = 3.0$, and $f_p = 4.0$ and slowly disappears below and above these values. It is also interesting to note that this dip occurs only at $B/B_\phi \approx 0.4$, corresponding to $d/d_\phi = 3/5$, where d and d_ϕ are the lattice constants of the initial vortex lattice and the pinning centers lattice, respectively. The size of pinning sites and high temperatures were found to have no effect on the dip effect.

In our simulations, the vortices were initially distributed in a square lattice over the whole system such that the number of vortices is less than the number of pinning centers. This arrangement gives a few rows of vortices initially located at or near the pinning centers. This initial distribution is mainly responsible for the formation of flow channels. On one hand, the rows of vortices which are initially located at or very close to the pinning centers get pinned with time and eventually form *pinned channels*. On the other hand, rows which are situated close to mid-distance between rows of pinning centers will likely form *flow channels*. The rest of vortices are randomly pinned. The vortex-vortex interaction modifies this picture slightly. The dynamics of vortices at the dip and away from it were investigated. After investigating this behavior further for different values of pinning strength f_p, we have found that the dip is always

24

associated with the formation of flow channels at $T = 0$ and $B/B_\phi \approx 0.4$. Figure 3 shows that F_d^c at the dip increases almost linearly with f_p.

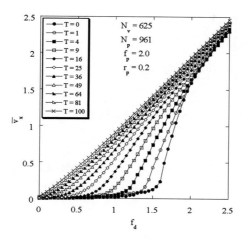

FIGURE 1. The average velocity $\overline{V}_x$ versus the driving force f_d calculated at different temperatures and for $N_p = 961$, $N_v = 625$ and $f_p = 2.0$.

CONCLUSION

We have carried out simulations on the behavior of the critical current density in high temperature superconductors with a very dense nanostructured periodic pinning sites. The variables in our simulations were the vortex density, the temperature, the pinning strength and the size of pinning sites. We have found that a dip in the critical current density occurs only at $T = 0$, $r_p = 0.2$ and $B/B_\phi \approx 0.4$ for different values of pinning strength f_p. Size of pinning sites and high temperatures were found to have no effect on the dip effect. The occurrence of such dip effect was attributed to formation of flowing and pinned channels of vortices.

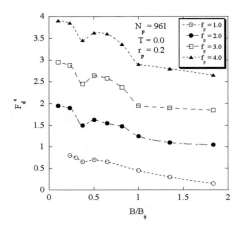

FIGURE 2. The critical depinning force F_d^c as a function of B/B_ϕ calculated for different values of pinning force f_p at zero temperature and a pinning radius $r_p = 0.2$.

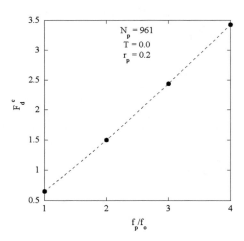

FIGURE 3. The critical depinning force F_d^c, at the dip as a function of the pinning strength f_p, at $T = 0.0$ for $r_p = 0.2$.

ACKNOWLEDGMENTS

This work was financially supported by the Research Affairs at the UAE University under a contract no. 03-02-2-11/06.

REFERENCES

1. G. Blatter, M. V. Feigl'man, V. B. Geshkenbein, A. I. Larkin, and V. M. Vinokur, Rev. Mod. Phys. **68**, 1125 (1994).
2. A. M. Campbell, and J. E. Evetts, Advancis in Physics **50**, 1249 (2001).
3. Y. Yeshurum, N. Bontemps, L. Burlachkov, and K. Kapitulnik, Phys. Rev. B **49**, 1548 (1994).
4. S. Anders, R. Parthasarathy, H. M. Jaeger, P. Guptasarma, D. Hinks, and R. G. v. Veen, Phys. Rev. B **58**, 6639 (1998).
5. M. Xu, T. W. Li, D. G. Hinks, G. W. Crabtree, H. M. Jaeger, and A. Haruyoshi, Phys. Rev. B **59**, 13632 (1999).
6. Y. Takahama, H. Suematsu, T. Matsushita, and H. Yamauchi, Physica C **338**, 115 (2000).
7. A. Otterlo, R. T. Scalettar, and G. T. Zimnyi, Phys. Rev. Lett. **84**, 2493 (1999).
8. C. J. Olson, C. Reichhardt, and S. Bhattacharya, Phys. Rev. B **64**, 024518 (2001); S. S. Banerjee, et al., Phys. Rev. B **62**, 11838 (2000).
9. M. Baert, et al., Phys. Rev. Lett. **74**, 3269 (1995); E. Rosseel, et. Al, Phys. Rev. B **53**, R2983 (1996).
10. C. Reichhardt, N. Gronbech-Jensen, Phys. Rev. B **63**, 054510 (2001).
11. I. M. Obaidat, U. Al Khawaja, M. Benkraouda, and N. Salmeen, Physics Letters A. **359**, 321 (2006); U. Al Khawaja, M. Benkraouda, I. M. Obaidat and S. Alneaimi, Physica C **442**, 1 (2006).
12. C. J. Reichhardt, G. T. Zimnyi, R. T. Scalettar, A. Hoffmann and Ivan K. Schuller, Phys. Rev. B **64**, 052503 (2001).
13. C. J. Reichhardt, G. T. Zimnyi and Niels Grnbech-Jensen, Phys. Rev. B **64**, 014501 (2001).
14. H. H. Solak, J. Phys. D: Appl. Phys. **39**, R171-R188 (2006).

Computing PI Polynomials of Some Nanostructures

H. Yousefi-Azari, B. Manoochehrian and A. R. Ashrafi

School of Mathematics, Statistics and Computer Science, University of Tehran, Tehran, Iran
Academy for Education, Culture and Research, Tehran, Iran
Institute of Nanoscience and Nanotechnology, University of Kashan, Kashan 87317-51167, Iran
hyousefi@ut.ac.ir

Abstract. A topological index is a numeric quantity that is mathematically derived in a direct and unambiguous manner from the structural graph of a molecule. It has been found that many properties of a chemical compound are closely related to some topological indices of its molecular graph. Among topological indices, the Wiener index is probably the most important one. This index was introduced by the chemist H. Wiener (1947) about 60 years ago to demonstrate correlations between physicochemical properties of organic compounds and the topological structure of their molecular graphs. He defined his index as the sum of distances between any two carbon atoms in the molecules, in terms of carbon-carbon bonds. The aim of this paper is to prepare some algorithms for computing distance matrix and Wiener index of zig-zag and armchair polyhex nanotubes.

Keywords: Distance matrix, Wiener index, zig-zag polyhex nanotube, armchair polyhex nanotube.
PACS: 81.07.De; 81.07.Nb; 81.16.Rf.

INTRODUCTION

Nanostructured materials have received a lot of attention because of their novel properties, which differ from those of the bulk materials. One-dimensional materials are an important category of nanostructured materials and have been widely researched yielding various special structures like nanotubes, nanorods, nanobelts and nanowires [1]. The materials in the nanotube form can be prepared from carbon.

Topological indices are numerical descriptors that are derived from molecular graphs of chemical compounds. Such indices based on the distances in a graph are widely used for establishing relationships between the structure of molecules and their physico-chemical properties [2,3]. One of the first of these was introduced early by Wiener [4]. He defined his index as the sum of distances between any two carbon atoms in the molecules, in terms of carbon-carbon bonds. Next Hosoya named such graph invariants, topological index [5]. With hundreds of topological indices one would expect that most molecules could be well characterized and their physicochemical properties correlated with the available descriptors. We encourage the reader to consult Refs. [6–8] and references therein, for further study on the topic.

In Refs. [7,8], the authors defined a new topological index and named it Padmakar-Ivan index. They abbreviated this new topological index as PI. This newly proposed topological index, PI, does not coincide with the Wiener index (W) for acyclic (trees) molecules. The derived PI index is very simple to calculate and has a discriminating power similar to that of the Wiener index, see [9]. This new index is defined as the sum of $[n_u(e|G) + n_v(e|G)]$ between all edges e=uv of a molecular graph G, i.e. $PI(G) = \sum_{e \in G}[n_{eu}(e|G) + n_{ev}(e|G)]$, where $n_u(e|G)$ is the number of edges of G lying closer to u than to v and $n_v(e|G)$ is the number of edges of G lying closer to v than to u.

The PI index for some classes of chemical graphs and nanotubes are computed in recent papers, [10-12]. The present authors also defined the notion of PI polynomial of a graph as an algebraic extension of its original definition. This polynomial is defined as $PI(G;x) = \sum_{\{u,v\} \subseteq V(G)} x^{N(u,v)}$, where

CP929, *Nanotechnology and Its Applications, First Sharjah International Conference*
edited by Y. I. Salamin, N. M. Hamdan, H. Al-Awadhi, N. M. Jisrawi, and N. Tabet
© 2007 American Institute of Physics 978-0-7354-0439-7/07/$23.00

$$N(u, v) = \begin{cases} n_{eu}(e \mid G) + n_{ev}(e \mid G) & uv \in E(G) \\ 0 & uv \notin E(G) \end{cases}, \tag{1}$$

see [13] for details.

Throughout this paper, our notation is standard and taken mainly from [2,3] and standard books of graph theory. We encourage reader to consult [10-13] and references therein for further study and background material on this topic.

MAIN RESULTS AND DISCUSSION

The aim of this section is computing the PI polynomial of some interesting class of nanostructures containing some types of graphen sheets and nanocrystals. To do this, we first prove a formula which is useful in our calculations. Let G be a molecular graph. Then we have:

$$PI(G; x) = \Sigma_{\{u,v\} \subseteq V} x^{N(u,v)}$$

$$= \Sigma_{uv \in E} x^{N(u,v)} + \Sigma_{uv \notin E} 1$$

$$= \Sigma_{e = uv \in E} x^{|E| - N(e)} + |E(K_n)| - |E| + |V| \tag{2}$$

$$= \Sigma_{e \in E} x^{|E| - N(e)} + \binom{|V| + 1}{2} - |E|.$$

Moreover, this formule shows that

(a) $PI'(G;1) = PI(G)$,
(b) $Deg\ PI(G;x) = |E(G)| - Min\{e \in E(G)\}\ N(e)$,
(c) The fixed term of PI $(G;x)$ is $\binom{|V(G)| + 1}{2} - |E(G)|$.

To prove (a), we notice that by definition of $N(u,v)$, $PI(G;x) = \Sigma_{u,v} x^{N(u,v)} = \Sigma_{\{u,v\} \in E(G)\}} x^{N(u,v)} + \Sigma_{(u,v) \notin E(G)} 1$. Therefore, $PI'(G;x) = \Sigma_{(u,v) \in E(G)} N(u,v) x^{N(u,v)-1}$. Substituting $x = 1$, we have $PI'(G;1) = \Sigma_{(u,v) \in E(G)} N(u,v) = PI(G)$.

To prove (b), we note that $Max(u,v) = Max(n_{eu} + n_{ev}) = Max\ (|E(G)| - N(e)) = |E(G)| - Min\{e \in E(G)\} N(e)$.

Finally, by definition, the fixed term of PI $(G;x)$ is the number of pairs (u,v) such that $N(u,v) = 0$. Such pairs have the form (u,u) or $\{u,v\}$, where u and v are non-adjacent edges of G. Since the number of pairs (u,u) is $|V(G)|$, $PI(G;0) =$ $|V(G)| + |E(\bar{G})| = \binom{|V(G)| + 1}{2} - |E(G)|$.

We now prove that if G is a connected graph with exactly m edges. Then $PI(G) \leq m(m-1)$ with equality if and only if G is a tree or a cycle of odd length. To prove this result, suppose $e = uv$ is an edge of G. Since $N(e) \geq 1$, $\Sigma_{e \in E(G)} N(e) \geq$ $\Sigma_{e \in E(G)} 1 = m$. Therefore, $PI(G) = m^2 - \Sigma_{e \in E(G)} N(e) \leq m^2 - m = m(m-1)$. We now assume that $PI(G) = m(m-1)$. If G is a tree then $PI(G) = (n-1)(n-2)$, as desired. So it is enough to consider non-acyclic graphs. Thus G has a cycle C of minimum length k, $k \geq 3$. If there exists an edge e for which $N(e) > 1$ then $\Sigma_{e \in E(G)} N(e) > m$ and so $PI(G) < m(m-1)$, which is a contradiction. Hence for every edge e, $N(e) = 1$. Suppose $C = x_1 x_2, x_2 x_3, \ldots, x_{k-1} x_k, x_k x_1$. We now consider two cases that k is odd or even.

Case 1. k is even. Suppose $f = x_1 x_2$ is an edge of C. Consider the edge $g = x_{\frac{k}{2}+1} x_{\frac{k}{2}+2}$. Since C has minimum length,

$$d(g, x_1) = d(g, x_2) = \frac{k}{2} - 1.$$

Thus, g is equidistant from both end of the edge f and hence $N(f) \geq 2$, a contradiction.

Case 2. k is odd. Suppose $f = x_1x_2$ and $v = x_{2+\frac{k-1}{2}}$. If deg(v) > 2 then we can choose an edge g = uv, in which u is distinct from x_i's. Hence $d(g,x_1) = d(g,x_2) = \dfrac{k-1}{2}$, which is impossible. Hence deg(x_1) = 2. Using a similar argument, we can see that for any i, $2 \le i \le k$, deg(x_i) = 2. But G is connected, so G = C, as desired. Conversely, if G is a tree or a cycle of odd length k then PI(G) = m(m−1), in which m = |E(G)|.

Apply these results to molecular graph of some nanostructures. We first consider the hexagonal graph of Figure 1, containing j hexagons in the jth row, $1 \le j \le n$. This graph is related to the atomic structure of bipod shaped nanocrystals, see [14, Figure 5]. Since the graph G has an equilateral figure, |E(G)|= 3(2 + 3 + ... + (n+1)) = 3/2(n^2 + 3n) and |V(G)| = 3 + 5 + ... + (2n+1) = n^2 + 4n + 1. Therefore,

$$
\begin{aligned}
PI(T(n); x) &= \sum_{\{u,v\} \subseteq V(T(n))} x^{N(u,v)} \\
&= \sum_{e \in E(T(n))} x^{|E(T(n))|-N(e)} + \binom{|V(T(n))|+1}{2} \\
&\quad - |E(T(n))| \\
&= 3\sum_{i=2}^{n+1} i x^{3/2(n^2+3n)-i} + \binom{n^2+4n+2}{2} \\
&\quad - 3/2(n^2+3n).
\end{aligned}
\tag{3}
$$

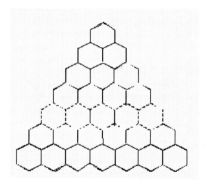

FIGURE 1. The Hexagonal Triangle Graph.

We next assume than H_n is an n-hexagonal net, which is a hexagonal system consisting of one central hexagon and is surrounded by n − 1 layers of hexagonal cells n ≥ 1, Figure 2. H_n is a molecular graph, corresponding to benzene (n = 1), coronene (n = 2) circumcoronene (n = 3), circum–circumcoronene (n = 4), etc. In [15], Shiu and Lam computed the Wiener index of an n-hegagonal net. They proved that W(H_n) = 1/5($164n^5$ − $30n^3$ + n). Here the PI polynomial of this graph is computed. Since the jth row of the graph H_n has exactly k + j vertical edges, |E(H_n)| = 3{2[(n+1) + (n+2) + ... + (2n − 1)] + 2n} = $9n^2$ − 3n. A similar calculation shows that |V(G)| = $6n^2$. So,

$$PI(H_n; x) = \sum_{\{u,v\} \subseteq V(H_n)} x^{N(u,v)}$$

$$= \sum_{e \in E(H_n)} x^{|E(H_n)| - N(e)}$$

$$+ \binom{|V(H_n)| + 1}{2} - |E(H_n)| \qquad (4)$$

$$= 3(2\sum_{i=1}^{n-1} (n+i)x^{9n^2 - 3n - (n+i)} + 2nx^{9n^2 - 5n})$$

$$+ \binom{6n^2 + 1}{2} - 9n^2 + 3n.$$

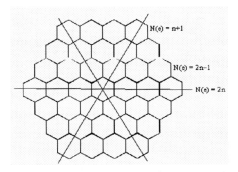

FIGURE 2. A 4-hexagonal net.

ACKNOWLEDGMENTS

The authors would like to thank Professor Mircea V. Diudea for giving us a copy of his software TopoCluj during the Math/Chem/Comp 2006 conference at Dubrovnik, Croatia. The second author was partially supported by a grant from IUT (CEAMA).

REFERENCES

1. B. O'Regan and M. Graetzel, Nature **353**, 737-740 (1991).
2. R. Todeschini and V. Consonni, Handbook of Molecular Descriptors (Wiley, Weinheim, 2000).
3. N. Trinajstic, Chemical Graph Theory (CRC Press, Boca Raton, FL. 1983).
4. H. Wiener, J. Am. Chem. Soc. **69**, 17-20 (1947).
5. H. Hosoya, Bull. Chem. Soc. Japan **44**, 2332-2339 (1971).
6. A. A. Dobrynin, Computers & Chemistry **23**, 43-48 (1999).
7. A. A. Dobrynin, R. Entringer and I. Gutman, Acta Appl. Math. **66**, 211-249 (2001).
8. A. A. Dobrynin, I. Gutman, S. Klavžar and P. Zigert., Acta Appl. Math. **72**, 247-294 (2002).
9. P. V. Khadikar, Nat. Acad. Sci. Lett. **23**, 113-118 (2000).
10. A. R. Ashrafi and A. Loghman, *MATCH Commun. Math. Comput. Chem.* **55**, 447-452 (2006).
11. A. R. Ashrafi and A. Loghman, J. Comput. Theor. Nanosci. **3**, 378-381 (2006).
12. A. R. Ashrafi and G.R. Vakili-Nezhad, Journal of Physics: Conference Series **29**, 181-184 (2006).
13. A. R. Ashrafi, B. Manoochehrian and H. Yousefi-Azari, Utilitas Mathematica, **71**, 97-108 (2006).
14. Y. -W. Jun, J.-W. Seo, S. J. Oh and J. Cheon, Coordination Chemistry Reviews **249**, 1766-1775 (2005).
15. W. C. Shiu and P. C. B. Lam, *Discrete Appl. Math.* **73**, 101-111 (1997).

Interconnect Challenges in the Nanometer Technology

H. A. S. Muhammad, S. K. Al Ahdab and A. Sagahyroon

Computer Engineering Department, American University of Sharjah, Sharjah, P.O. Box 26666, UAE
{asagahyroon}@aus.edu

Abstract. As technology sizes shrink, many new effects are being experienced by designers due to the use of nanometer technologies. At present, an integrated circuit chip has around 40 million transistors, and this number is expected to stretch to 4 billion by the year 2015 with no further scope for any reduction in the size. Deep submicron process technology will face with a number of new challenges in the area of testing, manufacturing, power consumption and specially the impact of interconnects (the wires linking transistors together) on performance. This paper examines the challenges incurred by interconnects in this deep-submicron era.

Keywords: Interconnects, crosstalk, signal integrity.
PACS: 85.35.–p

INTRODUCTION

Recently, the modern semiconductor industry has stepped into the nanometer range to implement and manufacture electronic circuits and SOCs (systems on chips) that have a broad range of applications. Conversely, the decreasing technology sizes and the increasing number of transistors per chip have lead to a multitude of challenges to be faced by the VLSI industry. The problems encompassing the interconnects in VLSI circuits include the increase in the propagation delay of transmitted signals and a significant decrease in the signal integrity. This paper discusses the reasons of the increase in the propagation delay due to crosstalk and also the decrease in the signal integrity as the supply voltage for the transistors is decreased [1-2].

The propagation delays in the transmission of signals between two components alter the time at which a certain signal reaches its destination; thus leading to numerous, unwanted logical errors. The major cause of these delays are the impedances(R and C) arising from the crosstalk effects (unwanted interference from other adjacent wires) and reflections in the interconnect wires. Moreover at extremely fast switching frequencies, there is a potential for inductance to play a vital role in the propagation delay. Furthermore, this increase in the impedance causes the quality of the propagating signal to deteriorate.

The drop in the signal integrity is also a result of the low voltage supply that leaves a very small range of voltages where the data can be interpreted correctly. In addition to that, the electromagnetic (EM) waves, which occur due to high current density, also interfere with the integrity of the signal. In other words, the signal suffers different levels of attenuations, resulting in an erroneous delivery of the data being transmitted by the interconnect [3-8].

MODELLING OF INTERCONNECTS

There are several possible models of interconnects, with different level of complexities starting from a simple lumped capacitor model to a frequency-dependent parameters (full wave) model. These models depend on the propagation of an electromagnetic (EM) wave in the complex metal-dielectric structure. Since the detailed EM analysis of interconnects required a large simulation time and complex computations, the EM models are therefore reduced to electrical models [4,6].

To appreciate the physics behind interconnects, two conductors with externally applied electric field between them are presented in Fig.2. These two conductors represent the voltage supply. Another conductor in the middle is the signal line which can be alternatively connected to the other two conductors by two ideal switches. In Fig.1.a, two different static charge and electric field distributions are resulting on the line (due to the effect of the external electric field). In Fig.1.b after the connection of the line to the upper conductor, the electric field is confined to the region between the lower metal and the line, whereas it is null in the region between the signal line and the upper metal. In 1.c, the field is confined in the region close to the upper metal. As a result of the switching event, an electromagnetic wave is generated, and is propagated in the dielectric between conductors [6].

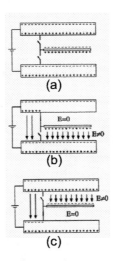

(a)

(b)

(c)

FIGURE 1. Ideal physical structure for describing a digital switching phenomenon

However, the interest lies in describing the phenomenon in terms of voltage and current at the connection points of each circuit element, rather than in terms of electromagnetic fields at each point in space.

In this paper, quasi TEM equations are assumed to model interconnects. Distributed RLC model is developed using these equations as shown in Fig. 2 [6-7].

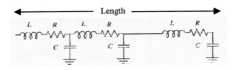

Figure 2. Distributed RLC model

EFFECTS OF CROSSTALK ON DELAY

Capacitive Coupling

Two conducting layers with a thin layer of a dielectric in between are generally used to create capacitance. In modern VLSI, the introduction of nanowires as interconnects has resulted in a new set of complexities that are associated with capacitance. The extremely close wires act as the conducting plates and the space between them acts as the dielectric material providing a perfect setting for the formation of capacitance. In the case of closely packed interconnects, capacitive coupling, defined as the intrusion of unwanted signals from neighboring wires, is produced. When signals in an adjacent wire change, they tend to affect the signal of the original wire causing it to become distorted. This phenomenon is known as crosstalk and is greatly influenced by the escalating aspect ratios.

Aspect Ratio (AR) is the thickness of a wire per unit width. The constantly decreasing width of the wires and the spacing between them has lead to a greater susceptibility to the creation of capacitance between wires resulting in a higher level of crosstalk.

The capacitive coupling takes place between two adjacent wires and metal layers as illustrated in Fig. 3 [3-4].

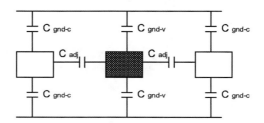

FIGURE 3. Coupling capacitance between adjacent wires and metal layers

When the signal in any of the wires changes, a certain amount of charge is transported to the coupling capacitor and a perturbation is created in the adjacent wire. The amount of charge that is transported to the coupling capacitance is equal to the adjacent capacitance times the voltage change that has occurred in the two wires [5].

$$\Delta Q = C_{adj} \Delta V \tag{1}$$

In the process of signal switching, the wire creating the disturbance is known as the culprit and the wire that is being affected is the victim [6]. A change in the culprit wire's signal creates an interference that adds to the propagation delay of the victim's signal. The change in the voltage of the victim is calculated as follows [5]:

$$\Delta V_{victim} = \frac{C_{adj}}{C_{gnd} + C_{adj}} \frac{1}{1+k} \Delta V_{culprit} \tag{2}$$

where k is defined as follows:

$$k = \frac{\tau_{culprit}}{\tau_{victim}} = \frac{R_{culprit}(C_{gnd-c} + C_{adj})}{R_{victim}(C_{gnd-v} + C_{adj})} \tag{3}$$

Fig. 1 shows all the capacitances that will be tolerated by each interconnect. Here every interconnect is surrounded by wires whose data switches frequently. The total capacitances seen by each interconnect in Fig. 1 is:

$$C_{TOT} = 2C_{gnd-v} + 2C_{adj} \qquad (4)$$

As explained in [3-4] and [6-7], the maximum effect of crosstalk is observed when the culprit and victim wires switch simultaneously but to opposite values (i.e. one of them switches to logic 0 and the other switches to logic 1). In this case, ΔV will be equal to $2V_{DD}$ and hence the coupling capacitor, C_{adj}, is considered twice as large. An experiment conducted in [6] showed the delay due to the switching of the signals in adjacent wires. The results of the experiment demonstrated that when three culprit wires, that were adjacent to a vitim wire, switched to a value opposite to that of the victim wire, a significant delay was created and the data from the input of the flip flop did not reach its destination in time.

References [3] and [8] discussed a model that analyzed the crosstalk capacitance. This model showed the link between two wires and explained the calculation of coupling capacitance while the signals in the wires are switching. In this model each pulling resistance consisted of the line resistance and the onchannel resistance linked with the line driver, where it was assumed that the NOT gates turns off immediately after the inputs are applied. The load capacitances, C_a and C_v consist of the line capacitance and the gate capacitance of the load driven by the line. Thus the line driver is equivalent to a resistance, and a network of coupling capacitors (C_m, C_a, C_v); Fig. 4 displays this model [8].

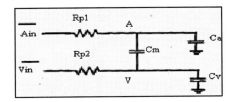

FIGURE 4: Capacitive coupling model

Inductive Coupling

Inductive crosstalk, defined as the transmission of noise through magnetic field to adjacent wires, has an effect on the delays of the signal at extremely high frequencies. In [9], the effects of inductive coupling are examined and the reasons stated for the increasing emergence of inductive coupling are the escalating frequencies and the lengths of the interconnects. At high frequencies, inductance L gives rise to considerable impedance that is equal to ωL.

At present, inductive coupling is becoming more significant with increasing frequencies and wire lengths. Wide wires are used in clock distribution networks and in upper metal layers [9-11]. Wider wires exhibit lower resistances and higher inductive effects.

Fig. 3 shows the regions where inductance has a more prominent effect on the coupling between wires. At upper metal layers, the value of the RC delay is smaller, thus the horizontal line in Fig. 5 is pushed upwards. As a result, the region in which the inductance is important increases.

Signal Integrity

Crosstalk has a larger impact on wires with larger resistance, as can be noted from Eq. (2). The higher the value of R_{victim}, the smaller will be the value of k and thus the greater the change in the value of V_{victim}. Subsequently, the resistance of the interconnects is also adding to the propagation delay of the signals. The particularly small width (in the nanometer range) of the interconnects has increased the value of the resistance tremendously. This can be proven by the classical equation of resistivity

$$R = \frac{\rho}{t}\frac{l}{w} \qquad (5)$$

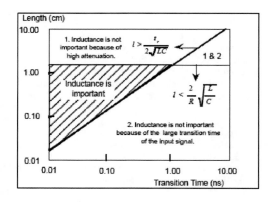

FIGURE 5: Transition time vs. Length of interconnect

where ρ is the resistivity of the metal that is used to construct the interconnects, l is the length, w is the width and t is the thickness of the interconnect.

The drop in signal integrity is another concern in the field of interconnects. As a result of the small wire geometries, the supply voltage is reduced to a value of 1V, with the threshold voltage being 0.2V [3][9]. Although this measure reduces the power dissipation in the wires, the amount of change that occurs in the signal due to crosstalk and other factors (not discussed in this paper) is very significant if compared to the supply voltage. Eventually, the signal reaching its destination is distorted to an extent that it cannot be interpreted correctly.

Consequently, the propagation delay that results from the coupling capacitors and the wire resistances adds to the other delays. This late delivery of data has a very profound effect in sequential circuits where the output in one stage depends heavily on the output from the previous stage.

CONCLUSION

In this paper, we discussed the effects of crosstalk on the propagation delay of signals. There are two different types of crosstalk that are imposing a concern in the VLSI industry. The first one is due to capacitive coupling and is increasing as the technology sizes have entered the nanometer range. The second type of crosstalk is due to inductive coupling and is a result of the increasing length of wires as the complexity of circuits is increasing. Consequently, the quality of the propagated signal, also known as signal integrity, is decreasing. The decrease in the supply voltage as a result of the extremely small dimensions of the transistors is also another cause of the possible deterioration in the signal integrity of VLSI designs.

REFERENCES

1. K.-T. Cheng, S. Dey, M. Rodgers, and K. Roy. "Test Challenges for Deep Sub-Micron Technologies" Proceedings of the 37th Design Automation Conference, 142-149, June, 2000.
2. M. T. Bohr, IEEE Transactions on Nanotechnology **1**, 56 (2002).
3. N. H. E. Weste and D. Harris, "Circuit Characterization and Performance Estimation," *CMOS VLSI Design*, Boston (Pearson Education, 2005) pp.196-216.
4. V. V. Deodhar and J. A. Davis, "Optimization of Throughput Performance for Low-Power VLSI Interconnects", *IEEE Transactions on VLSI Ssystems* **13** (2005).
5. R. Ho, K. Mai, and M. Horowitz, "The future of wires," *Proc. IEEE* **89** (2001) pp. 490-504.
6. F. Moll, "İnterconnection Noise in VLSI circuits" (Secaucus: Kluwer Academic Publishers, 2004).

7. C. N. Taylor, S. Dey and Y. Zhao, "Modeling and Minimization of Interconnect Energy Dissipation in Nanometer Technologies" DAC2001 (ACM, Las Vegas, Nevada, 2001).
8. W. Y. Chen, S. K. Gupta and M. A. Breuer, "Test Generation for Crosstalk-Induced Delay in Integrated Circuits," IEEE ITC International Test Conference, 1999.
9. Y. Ismail, E. Friedman, and J. Neves, "Figures of merit to characterize the importance of on-chip interconnect," *IEEE Trans. VLSI* **7**, 442 (1999).
10. J. Torres, "Advanced Copper Interconnections for Silicon CMOS Technologies," *Applied Surface Science* **91**, 112 (1995).
11. K. Likharev and V. K. Semenov, "RSFQ Technology for Sub-Terahertz-Clock Frequency Digital System," *IEEE Transactions on Applied Superconductivity* **AS-1**, 3 (1991).

Effect of Nano-Circular Inclusion on the Interfacial Stresses of a Nano-Composite

W. Kh. Ahmed, F. Kh. Omar and Y. Haik

Department of Mechanical Engineering, United Arab Emirates University, Al-Ain, United Arab Emirates
w.ahmed@uaeu.ac.ae

Abstract. The aim of this paper is to investigate numerically, through a finite element method, the influence of a Nano-Circular Inclusion on the interfacial stresses around a Nano-Fiber composite subjected to axial load to explain some of the causes that affect the failures of Nano-Composites. Previous investigations studied the effect of Nano-Fiber tip shape, Nano-Holes and its location around the Nano-Fiber on the interfacial stresses in a Nano-composite, while the current study focuses on studying the effect of the Nano-Circular Inclusion. It was assumed that Nano-Circular Inclusions were embedded in a discontinuous Nano-Fiber composite around a carbon Nano-Fiber composite. A tensile load was applied to the specimens in order to study the influence of these Nano-Circular Inclusions. It was found from the numerical analysis that Nano-Circular Inclusions have a great influence on the increase of the interfacial contact stresses around the Nano-Fiber in Nano-Composite.

Keywords: Nanofiber, Nanocircular Inclusion, Nanovoids, Interfacial stresses, Finite Element.
PACS: 81.07.–b

INTRODUCTION

Nanofibers and nanotubes have drawn huge attention from scientists and engineers worldwide because of their potential application in nano-scale polymer reinforcement. The focus on nano reinforcement composite, particularly carbon nano reinforcement composite CNF, has resulted in increasing attention to this newly promising material due to its extraordinary mechanical and electrical properties[1,2], particularly superior strength, stiffness and electrical and thermal conductivities. Researches have shown that carbon nanotubes exhibit exceptional mechanical properties [3]. Although there has been some variation in the reported values for the carbon nanotubes mechanical properties, the elastic modulus has been shown to be greater than 1 TPa, and the tensile strength exceeds that of steel by over an order of magnitude [4].

The tremendous mechanical properties of carbon nanotubes and other nano reinforcements can be realized only if efficient load transfer exists between the matrix and the reinforcement [5-8]. In some cases the load transfer between nanotubes and the surrounding matrix can be increased by introducing non-bonded interfacial compounds or chemical cross links between nanotubes and the matrix [9-12].

The stiffness properties of nanocomposites are always higher than those of the pure matrix; however, the final strength of the nanocomposite may or may not exceed the strength of the pure matrix if discontinuous nanofibers/nanotubes (even if they were aligned) are used in nanocomposites [13].

Many problems and challenges remain as barriers to the development and applications of nanomaterials including the development of techniques to produce nano-scale particles of high quality in sufficient quantities and at a low cost; the upgrade of the low fracture toughness and poor ductility of nanoscale materials, the assembly of nanocomponents into devices, and the improvement of the thermal stability of nanostructures [14].

CP929, *Nanotechnology and Its Applications, First Sharjah International Conference*
edited by Y. I. Salamin, N. M. Hamdan, H. Al-Awadhi, N. M. Jisrawi, and N. Tabet
© 2007 American Institute of Physics 978-0-7354-0439-7/07/$23.00

Another problem is the peeling of the matrix around the fiber if subjected to stresses due to the presence of nanovoids or nanocircular holes created through the manufacturing process. To achieve maximum utilization of the properties of nanofibers, uniform dispersion and good wetting of the nanofibers within the matrix must be ensured [15].

The local interfacial stress level in nanocomposites would be much higher than that in traditional composites because of high property mismatch between the nanoscale reinforcement and the matrix since high interfacial stress may lead to interfacial debonding and the final failure of nanocomposites, this may contribute to the low failure strains observed in nanocomposites [15,16]. In addition, the advantage of small diameters of nanofibers or nanotubes is an increased interfacial contact area with the matrix, while its shortcoming is a high possibility of initial interfacial defects, which may lead to low failure strain of nanocomposites.

The interfacial stress transfer and possible stress singularities, arising at the interfacial ends of discontinuous nanofibers embedded in a matrix, subjected to tensile and shear loading and the effects of Young's modulus and volume fractions on interfacial stress distributions, were investigated using finite element analysis [13] utilizing round-ended nanofibers to remove the interfacial singular stresses, which were caused by highly stiffness mismatch of the nanoscale reinforcement and the matrix. The normal stress induced in the nanofiber through interfacial stress transfer was still less than 2 times that in the matrix itself, this stress value is far below the high strength of the nanofiber. Therefore, the load transfer efficiency of discontinuous nanofibers or nanotube composites is very low [13].

Computational modeling techniques for the determination of mechanical properties of nanocomposites have proven to be very effective [17-24]. Computational modeling of polymer nanocomposite mechanical properties renders the flexibility of efficient parametric study of nanocomposites to facilitate the design and development of nanocomposite structures for engineering applications.

The finite element method (FEM) can be used for numerical computation of bulk properties based on the geometry and volume fractions of constituent phases [25-27]. FEM involves discretization of a material representative volume element (RVE) into elements for which the elastic solutions lead to determination of stress and strain fields. The coarseness of the discretization determines the accuracy of the solution. Nanoscale RVEs of different geometric shapes can be chosen for simulation of mechanical properties [17,18]. However, high complexity of models, expensive software, and time-consuming simulations limit the utility of this method. FEM-based micromechanics has been used extensively for the prediction of mechanical properties of nanostructured composites. Li et al. [28] used FEM-based approach to investigate the stress concentration at the end of carbon nanotubes and the effects of nanotube aspect ratio on the load transfer between nanotubes and matrix. Chen et al. [18] used different shapes of RVEs to understand the dependence of predicted properties on the element shape. Ahmed and Shakir [29] investigated using FEM the effect of noncircular nanoholes around the nanofiber on the interfacial stress in discontinuous nanofiber composites. They found that the nanoholes have a great influence in increasing the interfacial stresses around the nanofiber composite.

In this study, the influence of the proposed nanocircular inclusions resulting from the fabrication process around the nanofiber was investigated using finite element method. The Inclusions were assumed to be lied along the transverse and longitudinal sides of the fiber in addition to corner position one.

SIMULATION OF NANOFIBER COMPOSITES

Finite element analysis (FEA) was chosen as the primary tool for analysis instead of molecular dynamics simulations, since the latter could only deal with physical phenomena at the level of a few nanometers at the current stage, while the size of a representative volume of a nanocomposite material ranges from 10 nm to several hundreds of nanometers [30]. Since the smallest dimension of our nanofiber lies in the range 20–50 nm, continuum mechanics assumptions, such as those used in the finite element analysis, are still valid at such length scales [13]. Fisher et al.[30] and Chen and Liu [18] adopted similar finite element analyses with a focus on stiffness analysis incorporating micromechanics theory. It is noted that these finite element analyses simplified the complex interaction among the nanoscale reinforcement, matrix, and the possible interphase. The interphase issue has received considerable attention in nanocomposite systems, since nanoscale reinforcement affords a greater interphase volume compared to traditional composite materials. It is very difficult to determine the physical properties, such as thickness, of the interphase and they may certainly not be treated as material constants [13].Therefore, the concept of interphase and its modeling is not employed in this investigation.

Fiber end modifications techniques were effectively used to remove the interfacial stress singularity in macroscale dissimilar material joints through an integrated numerical and experimental investigation by Xu et al. [31]. Because the nanoscale experimental validation is very difficult, we mainly focus on numerical investigation. Although the nanofiber reinforced composite is the main focus, the numerical analysis can be easily extended to nanotube-reinforced composite by varying the stiffness and dimensions of the nanoscale reinforcement.

In this study, finite element analysis was used to investigate the influence of nanocircular Inclusions on the interfacial stresses in the RVE and the structural performance by utilizing (ANSYS5.4) finite element computer package [32]. The ANSYS finite element program is utilized to predict the interfacial stresses of RVE. The length and width of the RVE considered in this analysis is analogous to the L.Roy [13] and Ahmed [29].

In the finite element analysis, two dimensional (2D), 4-node solid element (Plane 42). Fig. 1 shows the dimension and the boundary conditions of the simulated RVE. It was attempted to maintain the same degree of refinement for all models to obtain consistent analysis.

The properties of the matrix are considered to be elastic isotropic. Matrix properties for Young's modulus and Poisson's ratio respectively are 2.6 GPa and 0.3. For the nanofiber, the properties 200GPa and 0.3 were used. The modulus of elastsity of the nanoinclusion considered as 1/100 of the matrix while 0.3 is adopted for the poisons' ratio.

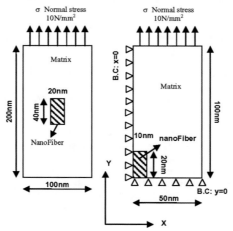

FIGURE 1. The dimensions and boundary condition of the RVE used for FEM.

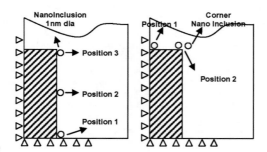

FIGURE 2. Nano-inclusion at the longitudinal (left), and transverse (right) edge of the nanofiber.

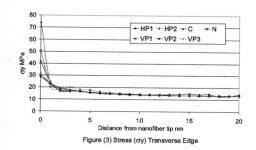

Figure (3) Stress (σy) Transverse Edge

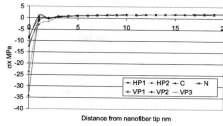

Figure (4) Stress (σx) at Longitudinal Edge

Two pairs of identical nanocircular Inclusions located symmetrically around the fiber in addition to a nonoiclusion at the corner of the nanofiber are shown in Fig. tensile load of 10 MPa stress was applied at the end of the RVE. Stresses at the short and long side of the specimen were measured.

RESULTS AND DISCUSSION

In the analysis of the RVE containing an Inclusion, the effect of the location of the nanocircular Inclusion along the fiber side, on the interfacial stresses, was studied, i.e., the traverse and the longitudinal side of the fiber. The results can be summarized as:

1. The corner inclusion increases σy along the nanofiber edge up to 1.8 times the stress of the non- inclusion case, while it increases the stress up to 2.83 times the σx stress along the longitudinal edge of the nanofiber, as shown in Figures (3) and (4).
2. It is evidence that the vertical position 3 near the tip of the longitudinal edge of the nanofiber increases the shear stress σxy up to 2 times and 1.2 times the stress of σxy for transverse and longitudinal edge of the nanofiber respectively, as shown in Figures(5) and (6).

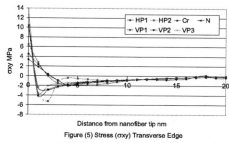

Figure (5) Stress (σxy) Transverse Edge

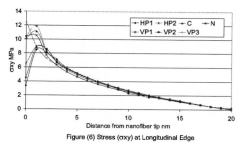

Figure (6) Stress (σxy) at Longitudinal Edge

HP=Horizontal Position, VP = Vertical Position, C = Corner Position, N =Non-Inclusion Case

CONCLUSIONS

This study is an attempt to explain the failure of the nanofiber composite in stresses below the proposed one via using FEM. It was shown in this study that the location of the nanocircular Inclusions around the nanofiber composite affects the rise up of the interfacial stresses many time compared with the nanocircular Inclusion-case for both transverse and longitudinal location of the nanocircular Inclusions.

REFERENCES

1. H. Ishikawa, S. Fudentani and M. Hrohashi, Mechanical properties of thin films measured by nanoundenters, App. Surf. Sci. **178**, 56-42 (2001).
2. B. Kracke and B. Damaschosile, Meaurement of nanohardness and nanoelasticity of thin gold films with scaning force microscope, Appl. Phys. Lett. **77**, 361-363 (2000).
3. M. M. J. Tracy, T. W. Ebbesen and J. M. Gibson, Nature **381**, 678 (1996).
4. P. K. Valavala and G. M. Odegrad, Modeling techniques for determination of mechanical properties of polymer nanocomposites, Rev. Adv. Mater. Sci. **9**, 34-44 (2005).
5. W. Huang, S. Taylor, K. Fu, Y. Lin, D. Zhang, T. Hanks, A. M. Rao, and Y. P. Sun, NanoLetters **311** (2002).
6. Velasco-Santos, A. L. Martinez- Hernandez, M. Lozada- Cassou and A. Alvarex- Castillo, Nanotechnology **13**, 495 (2002).
7. S. Banerjee and S. S. Wong, Journal of the American Chemical Society **124**, 8940 (2002).
8. S. B. Sinnott, Journal of Nanoscience and Nanotechnology **2**, 113 (2002).
9. S. J. V. Frankland, A. Caglar, D. W. Brenner and M. Greibel , J. Phys. Chem. B **106**, 3046 (2002).
10. Y. Hu, I. Jang and S. B. Sinnott, Composite Science and Technology **63**, 1663 (2003).
11. Y. Hu and S. B. Sinnott, Journal of Materials Chemistry **14**, 719 (2004).
12. G. M. Odegard, S. J. V. Frankland and T.S. Gates, in: AIAA/ ASME/ ASCE/ AHS Structures (Structural Dynamics and Materials Conference, Norfolk, Virginia, 2003).
13. L. Roy Xu And Sreeparna Sengupta, Interfacial Stress Transfer and Property Mismatch In Discontinuous Nanofiber and Nanotube Composite Materials, Journal of Nanoscience and Nanotechnology Vol. X, 1– 6, 2005.
14. The Impact of Materials: From Research to Manufacturing (National Academies Press, Washington, 2003) http://www.nationalacademies.org.nmab.
15. W. H. Zhong, J. Li, L. R. Xu, J. A. Michel, L. M. Sullivan, and C. M. Lukehart, *J. Nanosci. Nanotech.* **4**, 794-802 (2004).
16. L. R. Xu, V. Bhamidipati, W. H. Zhong, J. Li, C. M. Lukehart, E. Lara- Curzio, K. C. Liu, and M. J. Lance, *J. Comp. Mater.* **38**, 1563 (2004).
17. Y. J. Liu and X. L. Chen, Mechanics of Materials **69**, 35 (2003).
18. X. L. Chen and Y. J. Liu, Computational Materials Science **29**, 1 (2004).
19. Y. Liu, N. Nishimura and Y. Otani, Computational Material Science (in press).
20. K. Van Workum and J. J. De Pablo, Nano Letters **3**, 1405 (2003).
21. S. A. Ospina, J. Restrepo and B. L. Lopez, Materials Research Innovations **7**, 27 (2003).
22. N. Sheng, M. C. Boyce, D. M.Parks et al., Polymer **45**, 487 (2004).
23. T. M. Gates and J. A. Hinkley, NASA, TM- 2003- 212163.
24. G. M. Odegard, T. S. Gates, K. E. Wise, C. Park and E. J. Siochi, Composites Science and Technology **63**, 1671 (2003).
25. D. F. Adams, Journal of Composite Materials **4**, 310 (1970).
26. T. H. Lin, D. Salinas and Y. M.Ito, Journal of Composite Materials **6**, 48 (1972).
27. G. J. Dvorak, M. S. M. Rao and J. Q. Tarn, Journal of Composite Materials **7**, 194 (1973).
28. C. Y. Li and T. W. Chou, Journal of Nanoscience and Nanotechnology **3**, 423 (2003).
29. W. Kh. Ahmed, S. A. Shakir," The Influence of Nanoholes on the Interfacial Stresses in Discontinuous Nanofiber Composite", International Conference on Bio-Nanotechnology: November 18-21, 2006, Al-Ain, United Arab Emirates.
30. F. T. Fisher, R. D. Bradsha, and L. C. Brinson, *Comp.Sci. Technol.* **63**, 1689 (2003).
31. L. R. Xu, H. Kuai, and S. Sengupta, Experimental Mechanics, **44**, 608-615 (2004).
32. S. Moaveni" Finite Element Analysis, theory and application with ANSYS" (Prentice Hall Inc., 1999).

Small Nanoclusters of ZnO and ZnS: a Density-Functional Study

A. AlSunaidi

Physics Department, King Fahd University of Petroleum and Minerals, Dhahran 31261, Saudi Arabia
asunaidi@kfupm.edu.sa

Abstract. We report the results of a comparative study of the properties of small $(ZnO)_n$ and $(ZnS)_n$ nanoclusters with $2 \le n \le 7$. Planar rings were found to be the most stable structures for the ZnO. For ZnS, planar rings were found for $n = 2, 3$ and 4 and a non-planar ring for $n = 5$. The most stable structures for $n = 6$ was a drum-like and the $n = 7$ was a chair-like structure. The nucleation energy of the clusters showed that clusters made of hexagons ($n = 3$) are more stable than other clusters. The HOMO-LUMO gap energy is found to be greater for the smaller cluasters than the bulk value, but highly dependent on the structure.

Keywords: Nanoparticles, Zinc-Oxide, Zinc-Sulfide, Density-Functional Theory.
PACS: 36.40.Cg, 61.46.Df.

INTRODUCTION

Zinc-oxide (ZnO) and zinc-sulfide (ZnS) are II–VI semiconductors that are of interest because of their wide range technological applications. For example, these compound semiconductors are used as catalysts [1], solar cells [2] and sensors [3]. Bulk ZnO exists in the hexagonal wurtzite structure. It has a wide bandgap of 3.3 eV with a relatively high exciton binding energy (60 meV). In particular, ZnO microstructures have attracted increasing attention due to their UV lasing [4] and optoelectronic properties [5]. There are two phases of ZnS: cubic zincblende and hexagonal wurtzite. Zincblende is slightly more stable than wurtzite. The cubic form has a band gap of 3.54 eV at 300 K whereas the hexagonal form has a band gap of 3.91 eV.

In recent years, there has been much experimental work designed to study the properties of metal oxide nano-structures. Different techniques were used to synthesize ZnO and ZnS nanoclusters. For example, stoichiometric and non-stiochemetric ZnO nano-clusters were produced experimentally using the laser ablation technique [6]. Positively charged zinc oxide clusters $(ZnO)_n$ (up to $n = 16$) of various stoichiometry were synthesized in the gas phase when a fresh ZnO surface is irradiated by excimer ArF laser. The same technique was used by Burnin and BelBruno [7] who noticed abundance of $(ZnS)_{13}$ in the mass spectrum. Also, small $(ZnO)_n$, $n = 1-15$, nanoclusters (of size in the range 0.3-1.0 nm) were prepared using the elecroporation of unilamellar vesicles [8] and their optical properties were in agreement with theoretical predictions. Larger ZnO nanoparticles of size 7 nm were synthesized [9] by sputtering a Zn target with Ar+ ions into an atmosphere of Ar, He and O_2 gases. The Zn atoms coalesce with oxygen atoms and Ar atoms due to the three-body collisions to form ZnO nanoclusters.

Nanolclusters of ZnO and ZnS were also studied using computer simulation. Simulated annealing (molecular dynamics) simulations were carried out by Behrman et al. [10] to study the stability of spheroid structures of ZnO using a simple interatomic potential of the Born-Mayer type. Their molecular dynamics simulations, supported by quantum molecular orbital calculations suggested that $(ZnO)_n$ with $n = 11, 12$ and 15 form spheroids as the most stable structures. Also, ring-forming clusters of $(ZnO)_n$, with $n < 7$ were investigated [11] using ab initio calculations using

CP929, Nanotechnology and Its Applications, First Sharjah International Conference
edited by Y. I. Salamin, N. M. Hamdan, H. Al-Awadhi, N. M. Jisrawi, and N. Tabet
© 2007 American Institute of Physics 978-0-7354-0439-7/07/$23.00

ultrasoft pseudopotentials and a plane-wave basis set. For each value of n they assumed several possible structures and found that the ring structure is always the global minimum. Matxain et al. have used density-functional theory to study the structures of small ZnO [12] and ZnS [13] nanolcusters. The geometries were optimized using the B3LYP approximate gradient-corrected DFT with the SKBJ basis set. For ZnO, ring-like structure were found for clusters of sizes $2 \leq n \leq 7$, while for ZnS only clusters in the range $2 \leq n \leq 5$ exhibited ring-like structures.

Recently, we made calculations based on a genetic algorithm to predict all the possible local minimum structures of $(ZnO)_n$ [14]. We were able to get clusters of sizes up to $n = 30$. In this work, we intend to make a comparison between the properties of small nanoclusters ($2 \leq n \leq 7$) of ZnO and ZnS using density functional theory. Although nanoclusters of these two compounds were studied before, but here we make better estimates of the energies and structures using better basis sets. In addition, we will compare the structures including angles and bond lengths as well as the nucleation and HUMO-LUMO gap energies. This paper is organized as follows: The next section will give a brief description of the computational methods used. Then the results will be presented and discussed in section III. Finally a summery of the results will be presented in the conclusion.

COMPUTATIONAL METHOD

Elsewhere [14], we have found the most stable local minimum structures for the ZnO clusters using a hybrid Genetic algorithm. The interatomic potentials used were combination of a Coulomb potential, a Buckingham potential, a Lennard-Jones potential and a power-series potential. A potential representing the polarization of the oxygen atoms was also included. In this work we have used the Gaussian 03 package [15] to locate the global minimum out of the many local minima found in our previous work. The computations were performed employing the density-functional theory (DFT) at the B3LYP level. The basis sets used were the TZVP (triple-ξ valence plus polarization set) augmented with diffuse s and p functions. This combination is known to give better estimates of the energy compared to those used in previous studies. We have optimized all the local minima found for $(ZnO)_n$ using the DFT. For each value of n, the one with the lowest energy is considered to be the global minimum. For $(ZnS)_n$, we have used the same local minima configurations found for $(ZnO)_n$. We then scaled the bond lengths to a value of 2.1 $\overset{\circ}{A}$, close to the expected bond length for ZnS. Then we optimized the clusters using the DFT. As for $(ZnO)_n$, the ones with the lowest energy are considered the global minimum.

RESULTS

The global minimum (GM) structures for the $(ZnO)n$ clusters of sizes $2 \leq n \leq 7$ are shown in Fig. 1. This is the set of ring-forming clusters. Although our Genetic algorithm (GA) [14] have produced several different structures (isomers) for the same n, we selected to show only those which were found to be the global minimum by DFT. The square for $n = 2$ and the hexagon for $n = 3$ are believed to be the basic building block for larger clusters. Linear structures for these cluster sizes (not shown) were found to be less stable. For $n = 4$, a ring was also found to be the GM but the GA has produced also cubic structure (rhombic) but is less stable than the ring shown in the figure. The $n = 5, 6$ and 7 cluster also has the ring structure as the lowest energy structure.

It is important to know how the bond length and the Zn-O-Zn and the O-Zn-O angles change with n. The bond length of the ZnO unit is very close to that in the ZnO bulk wurtzite lattice. However, for $(ZnO)_2$, the bond length is close to 0.19 nm, but decreases as n increases to reach a value close to that of the bulk when $n = 7$. We notice also that the Zn-O-Zn angle increases with n to reach a value close to 180° at $n = 7$, while the O-Zn-O angle is always smaller than the Zn-O-Zn angle.

For ZnS, the global minimum structures, as shown in Figure 2, for $2 \leq n \leq 4$ are planar rings. The $n = 5$ cluster exhibits a non-planar ring with four sulfur atoms lie in a plane but the fifth atom along with two zinc atoms lies out of the plane. A drum-like structure made of two hexagons is found for $n = 6$. The hexagons in this case are not planar. The global minimum structure for $n = 7$, is a chair-like structure made of $n = 3$ and $n = 4$ non-planar rings. The ring-structures for the $5 \leq n \leq 7$ ZnS clusters are not global minima as in the ZnO.

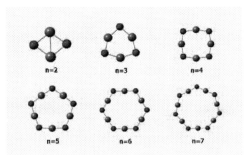

FIGURE 1. (Color online) Global minimum structures for the ZnO nanoclusters. Zn in blue and O in red.

Investigating the bond lengths, we notice that for the ZnS clusters, the bond length is always greater than that of the ZnO clusters. It decreases as n increases from 2 up to 5. The same increase is seen in the angles. For $n = 5$, the bond length is alternating between 0.1177 and 0.1181 nm. The Zn-S-Zn angle has its lowest value ($94°$) at the out-of-plane sulfur atom. Also, the out-of-plane Zn atoms exhibit the maximum S-Zn-S angles ($178°$). For the $n = 6$ cluster, atoms within both hexagons have equal bond lengths and Zn-S-Zn and S-Zn-S angles. The bond length between the atoms belonging to different hexagons is, however, greater. The bond length between atoms in the $n = 7$ cluster varies and the S atom at the top makes the shortest length of 0.2149 nm. The angles within the $n = 3$ and $n = 4$ rings (making the $n = 7$ cluster) also vary, and the S atom at the top makes the smallest Zn-S-Zn angle of $135°$.

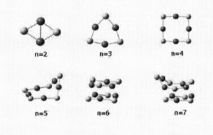

FIGURE 2. (Color online) Global minimum structures for the ZnS nanoclusters. Zn in blue and S in yellow.

Stability of the clusters is usually investigated by calculating the evolution of the nucleation energy defined here as

$$E_{nuc}(n) = E(n) - E(n-1) - E(1) \tag{1}$$

where $E(n)$ is the total energy of the (ZnO)$_n$ or the (ZnS)$_n$ cluster and $E(1)$ is the energy for a ZnO or ZnS unit. This is the energy needed to grow a cluster n from the cluster ($n-1$). The more negative the nucleation energy for a given cluster, the more stable it becomes. Figure 4 shows the nucleation energies of the (ZnO)$_n$ and The (ZnS)$_n$ clusters. It is clear that for both ZnO and ZnS, the hexagons ($n = 3$) are the most stable. For ZnO, it seems that as the ring size becomes larger the cluster becomes more unstable. For the ZnS, the drum structure for the $n = 6$, is more stable that the $n = 5$ and $n = 7$ structures because it is made of two $n = 3$ hexagons.

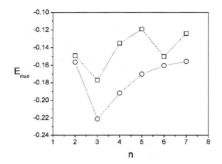

FIGURE 3. Stability of the ZnO (circle) and ZnS (square) nanoclusters represented as the nucleation energy as a function of cluster size.

Many electrical and optical properties are related directly to the band gap. We have calculated the band gap as the energy difference between the highest occupied (HOMO) and the lowest unoccupied (LUMO) molecular orbitals. Figure 2 shows the band gap energies for the ZnO and the ZnS clusters. Remember that the band gap for the ZnO is 3.3 eV and for the ZnS is 3.54. Once the clusters have the hexagon structure ($n = 3$), the band gap gets a value larger than its bulk value. The band gap energies for the ring structures for ZnO and ZnS are almost constant and equal. The three-dimensional structures for $n = 6$ and 7 of the ZnS have lower band gap energies than the rings.

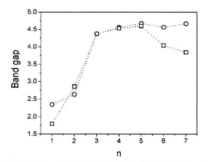

FIGURE 4. The HOMO-LUMO band gap for the ZnO (circle) and ZnS (square) as a function of cluster size n.

CONCLUSION

In this work we have compared several properties on the ZnO and ZnS nanoclusters of size $2 \leq n \leq 7$. Our first observation was that three dimensional structure which has a coordination number of more than two are more preferable for ZnS once the value of $n = 6$. In fact, even for $n = 5$, the clusters starts showing deviation from the more symmetric planar ring-like structure. For ZnO, the ring-structure is always preferred for these clusters sizes. The influence of these differences in the structure is reflected on the stability and band gap energies. We have noticed that the $n = 6$ drum made of two hexagons, is more stable than the ring structure of ZnO $n = 6$. We have also seen that the band gap energy depends strongly on the geometry of the structures. The ring-like structures of both ZnO and ZnS with the same n, have

46

the same band gap energies. However, the three-dimensional structures of ZnS have lower band gap energies. This could be related to the increase in the coordination number.

ACKNOWLEDGMENTS

The author acknowledges support from King Fahd University of Petroleum and Minerals (KFUPM) and King Abdulaziz City for Science and Technology (KACST).

REFERENCES

1. M. Hoffmann et al, *Chem. Rev.* **95**, 69 (1995).
2. A. Ennaoui, M. Weber, R. Scheer and H. Lewerenz, *Sol. Ener. Mat. Sol. Cells* **54**, 277 (1998).
3. P. Tran, E. Goldman, G. Anderson, J. Mauro, H. Mattawsi, *Phys. Stat. Solidi B* **299**, 427 (2002).
4. F. H. Nicoll, *Appl. Phys. Lett.* **69**, 13 (1996).
5. D. M. Bagnall, Y.F. Chen, Z. Zhu, T. Yao, S. Koyama, M. Y. Shen and T. Goto, *Appl. Phys. Letters* **70**, 2230 (1997).
6. U. Ozgur, Ya. I. ALivov, C. Liu, A. Teke, M. A. Reshchikov, S. Dogan, V. Avrution, S.-J. Cho, and H. Morkoc, *J. Appl. Phys.* **98**, 041301 (2005).
7. A. Burnin and J. BelBruno, *Chem. Phys. Letters* **361**, 341 (2002).
8. S. Wu, N. Yuan, H. Xu, X. Wang and Z. Nanotechmology **17** 4713 (2006).
9. J. Antony, X. B. Chen, J. Morrison, L. Bergman, and Y. Qiang D. E. McCready and M. H. Engelhard, *Appl. Phys. Letters* **87**, 241917 (2005).
10. E. C. Behrman, R. K. Foehrweiser, J. R. Myers, B. R. French, M. E. Zandler, *Appl. Phys. Rev. A* **49**, R1543 (1994).
11. A. Jain, V. Kumar, and Y. Kawazoe, *Comp. Mat. Sci.* **36**, 258 (2006).
12. J. M. Matxain, J. E. Fowler, J. M. Ugalde, *Phys. Rev. A* **62**, 053201 (2000).
13. J. M. Matxain, J. E. Fowler, J. M. Ugalde, *Phys. Rev. A* **63**, 013202 (2000).
14. A. AlSunaidi, A. Sokol, C. R. Catlow and S. Woodley, *in preparation.*
15. M. J. Frisch et al., *Gaussian 03, Revision B.0.3,* 2003 (Gaussian Inc.: Pittsburg, PA).

QUANTUM WELLS, QUANTUM WIRES,
AND QUANTUM DOTS

Role of Potential Steps on the Rashba Effect in Quantum Wells

S. Lamari

Department of Physics, University of Ferhat Abbas, Setif, Algeria
s_lamari@yahoo.fr

Abstract. Using our recent effective mass theory of the Rashba effect [S. Lamari, Phys. Rev. B 75, in press] we explore the modifications in the strength of the Rashba spin splitting in GaAs-AlGaAs quantum wells. It is shown that adding potential steps reduces the spin splitting.

Keywords: Rashba effect, quantum wells, band offset, GaAs.
PACS: 73.40.Qv; 71.70.Ej; 73.61.Ey; 73.21.-b

INTRODUCTION

Following the seminal paper of Datta and Das[1] where a new device called the spin FET was proposed, the study of of spin orbit effects in semiconductor research has increased tremendously[2]. In this new device spin polarized electrons are injected by a ferromagnetic source and collected by a ferromagnetic drain. If the length of the conduction channel is smaller than the length of spin dephasing through Dyakonov Perel mechanism[3] then transport is ballistic and a phase shift $\Delta\theta$ appears between spin components. With m^* being the effective mass and α the Rashba parameter[4] this phase shift is $2\,m^*\alpha L$. The same authors then predicted spin modulation of current by a gate. This outstanding idea was greeted with much enthusiasm and motivated a large number of experiments worldwide[5]. Nonparabolicity[6,7] and population[7] of subbands turned out essential. Our work in Ref. 7 was numerical, but recently[8] we introduced a fully analytic theory of the Rashba parameter using the invariant expansion of Rossler *et al*[9]. The present contribution is a numerical implementation of this framework.

INTRODUCTION

Let z denote the growth direction and $V(z)$ the external potential felt by an electron then

$$V(z)=V_0(z)+v(z), \tag{1}$$

where V_0 and v are, respectively, the potentials due to band offsets and charges. In the absence of SO, for (001) grown heterostructures the electron wavefunction satisfies the equation[8]

$$(A\hat{k}_z^2+C+B\hat{k}_z^4+V)\,\zeta_n(z)=E_n\zeta_n(z), \tag{2}$$

where

$$A = a_{12} + (2a_{13} + a_{14})k^2, \tag{3}$$

$$B = a_{13}, \tag{4}$$

$$C = a_{12}k^2 + \left(a_{13} + a_{14}\frac{1-\cos(4\theta)}{8}\right)k^4. \tag{5}$$

CP929, *Nanotechnology and Its Applications, First Sharjah International Conference*
edited by Y. I. Salamin, N. M. Hamdan, H. Al-Awadhi, N. M. Jisrawi, and N. Tabet
© 2007 American Institute of Physics 978-0-7354-0439-7/07/$23.00

In these equations k is the wavevector along the interface. With a_{64} the bulk spin orbit constant[9] the Rashba parameter is[8]

$$\alpha = \left| \frac{1}{e} \left\langle \frac{dv}{dz} \right\rangle_n \right| . \tag{6}$$

If we neglect nonparabolicity and anisotropy by setting a_{14} and a_{13} to 0 then

$$\alpha = \frac{a_{64}}{e} \sum_{i=1}^{N} P_i \, \Delta V_i \ . \tag{7}$$

The sum is over the N interfaces of the structure and P_i, ΔV_i the probability and potential offset at the ith interface. The self-consistent potential is obtained from the Poisson equation

$$\frac{d^2 v}{dz^2} = -\frac{4\pi}{\varepsilon} e \rho(z) , \tag{8}$$

with $\rho(z) = \rho_{ion} + \rho_e$ and

$$\rho_e = -e \sum_{occupied} \zeta_n^2(z) . \tag{9}$$

NUMERICAL RESULTS

We used Eqs. (1) – (6) to explore the Rashba spin splitting in GaAs-AlGaAs quantum wells. For simplicity in our calculations drop the effects of nonparabolicity and anisotropy, moreover apart from the band gaps we also take the same values for the band structure parameters throughout the heterostructure. The quantum wells for which we carry out simulations have no external gates, and are doped uniformly in the narrow region z = -2nm—z=-1nm, with z=0 corresponding to the left interface. The electron density is deduced from charge neutrality. Fig.1 shows a typical self-consistent potential. In the present case only two bound states exist, but electrons populate only one subband because of the large density of states of GaAs.

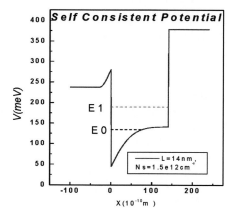

FIGURE 1. Self-consistent potential and energies of two bound states.

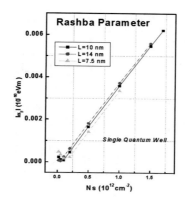

FIGURE 2. Absolute value of α as a function of the 2d electron density.

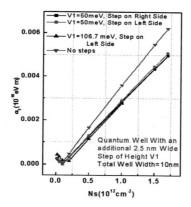

FIGURE 3. Rashba parameter for a stepped well.

In Fig. 2 we show as a function of Ns the influence of the well width on α where we see that this parameter decreases with well width at large values of Ns but have a different behavior at low values. This dependence can be understood using Eq. (7).

Fig. 3 shows the influence of an additional 2.5 wide step of height V_1 located on either side within the quantum well. One immediately notes that adding a step brings down the Rashba parameter. In Fig. 4 we show the probabilities at the interfaces of quantum wells of different widths as functions of the 2d electron density Ns. It is interesting to note that for narrow wells the curves representing the probabilities at either interface cross. This means that in this case the Rashba parameter vanishes.

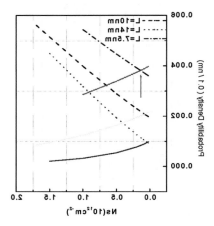

FIGURE 4. Probability density at the interfaces of single quantum well as a function of 2d electron density. The dashed curves are for the left interface while the solid lines are for the right interface.

CONCLUSIONS

Through extensive self-consistent calculations performed in the Hartree approximation we have investigated numerically the behavior of the Rashba parameter in GaAs-GaAlAs quantum wells as a function of the 2d electron density Ns. On the basis of our recent effective mass theory of the Rashba parameter[8] we showed that addition of steps can modify the strength of the Rashba effect. Moreover, for a 7.5 nm wide well for instance, the Rashba effect is predicted to vanish at $Ns=0.154\times10^{12}$ cm^{-2}.

ACKNOWLEDGMENTS

I wish to thank the Algerian Ministry of Higher Education for support and the conference organizers for financial help.

REFERENCES

1. S. Datta and B. Das, Appl. Phys. Lett. **56**, 665 (1990).
2. I. Zutic, J. Fabian, S. Das Sarma, Rev. Mod. Phys.**76**, 323 (2004).
3. G.E. Pikus and A. N. Titkov, in Optical Orientation, edited by F. Meier and B. P. Zakharchenya (North Holland, Amsterdam, 1984).
4. Y. A. Bychkov and E.I. Rashba, J. Phys. C **17**, 6039 (1984).
5. J. Nitta et al, Phys. Rev. Lett. **78**, 1335 (1997); Can Ming Hu et al., Phys. Rev. B **60**, 7736 (1999); A. C. H. Rowe et al., ibid **63**, 2013307R (2001).
6. T. Matsuyama et al., Phys. Rev. B **61**, 15588 (2000); C. Schierholz et al., ibid **70**, 233311 (2004).
7. S. Lamari, Phys. Rev. B 67, 165329 (2003); Phys. Rev. B **64**, 245340 (2001).
8. S. Lamari, Phys. Rev. B **75**, in press.
9. G. Lommer et al., Phys. Rev. B **32**, R6965 (1985); M. Braun et al., J. Phys. C **18**, 3365(1985); H. Mayer et al., Phys. Rev. B **44**, 9048 (1991).

3D Self-Organization of InGaAs Quantum Dots Grown Over GaAs <001>

M. L. Hussein[1], E. Marega Jr.[2], and G. J. Salamo[2]

[1] Applied Physics Department, Tafila Technical University, Jordan
[2] Microelectronics-Photonics, University of Arkansas, Fayetteville, AR 72701, USA
muhamad197@yahoo.com

Abstract. Lateral and vertical ordering of InGaAs quantum dots over GaAs <001> has been achieved in earlier reports resembling an anisotropic lateral pattern. We present in this letter a method of breaking the anisotropy of such structures by changing the growth environment from As_4 to As_2 molecules, a method that made it possible to produce 2D lateral ordering of InGaAs QDs. Our results are consistent with reported experimental and theoretical studies on surface structure and diffusion mechanism over GaAs <001>.

Keywords: Self-organized, quantum dots, lateral ordering, vertical stacking, molecular beam epitaxy.
PACS: 81.07.-b, 81.07.Ta

INTRODUCTION

III-V quantum dots (QDs) have been investigated as a promising candidate for a new generation of lasers, detectors and photonic crystals [1]. However, many applications require uniformity in size and shape and control of the spatial distribution of the QDs in order to realize their potential advantages [1,2]. For example, application of InGaAs QDs grown on GaAs <100> has been limited due to the random nucleation of the self-assembly process [3]. While recent approaches using lithographic techniques have demonstrated successful lateral ordering of QDs [4-6], these methods require extra processing steps which also might introduce defects that prevent successful application.

In an effort to overcome these difficulties, self-organization of InGaAs QDs on a GaAs <001> surface has been recently investigated through vertical stacking of QDs with growth at relatively high temperatures [7] and yielded "chains" of ordered QDs that was explained in terms of preferred adatom diffusion and strain relaxation along the <0-11> direction [8-9]. The chain-like lateral ordering is characterized by QDs that are very close to each other along the chain direction, making the 2D lateral ordering on GaAs (001) surface difficult to achieve due to the anisotropy of that surface [10]. High index substrates, on the other hand, have been used to engineer a uniform surface diffusion profile to achieve vertically ordered planes of 2D laterally ordered QD [11]. While this approach has been successful, for expensive substrates, one can expect a preference for lateral ordering on <001> substrates.

In this paper, we report on the observation of 2D laterally ordered self-assembled QDs with low indium composition (less than 40%) on a GaAs <001> surface. This was accomplished by using As_2 molecules as opposed to As_4 molecules as a background environment to dramatically alter the diffusion profile of adatoms and the corresponding strain profile during the formation of QDs.

CP929, *Nanotechnology and Its Applications, First Sharjah International Conference*
edited by Y. I. Salamin, N. M. Hamdan, H. Al-Awadhi, N. M. Jisrawi, and N. Tabet
© 2007 American Institute of Physics 978-0-7354-0439-7/07/$23.00

EXPERIMENT

The InGaAs QDs investigated in this work were grown by molecular beam epitaxy (MBE) on semi-insulating GaAs <001> substrates. Following a 0.3 μm GaAs buffer layer grown at 580° C to smooth the surface, 16 layers of InGaAs QDs spaced by 60 mono-layers of GaAs were grown at 540° C. The last QD layer was left uncapped for morphology analysis. Two types of samples were grown, one under As_4 and the other under As_2 background environment with a beam equivalent pressure of V/III ratio of 15:1 in both cases. As_2 molecules were produced by cracking As_4 molecules at 900° C. The QDs growth was monitored by in situ reflection high-energy electron diffraction (RHEED). Surface morphologies of the samples were characterized by atomic force microscopy (AFM).

RESULTS AND DISCUSSION

Diffusion Factor

The AFM images in Figure 1(a) and (b) represent the morphology of the last layer of InGaAs QDs samples grown under As_4 and As_2 environment respectively, with indium composition of 40% and nominal deposition of 8 mono-layers chosen to keep the deposition at 20% above the critical thickness. The image in Figure 1(a) reflects the well known QD chains structure, in which ordering occur only along the <0-11> direction due to anisotropic strain field transfer through GaAs spacers [8, 9]. The average QD height is 9.5 nm, however, the height modulation along the chains is less than 3 nm. In this case the dots resemble a wire-like structure. On the other hand, for sample (b) grown under an As_2 flux, the spacing between the centers of dots along the chain is larger than the dot diameter resulting in QDs that are more round-like and well separated from each other. Comparing the average dot height and diameter under As_4 (9.5 nm and 62 nm respectively) with those under As_2 (12 nm and 50 nm respectively) reveals the As_2 restriction on the dot to spread out in diameter which has been compensated by an increase in the dot height. Figure 1(b) reveals the amazing effect of As_2 to break up the anisotropy in lateral ordering, thus opening the potential for achieving a 2D ordering of QDs.

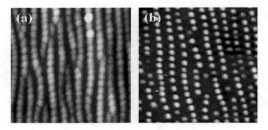

FIGURE 1. 1×1 micron AFM top views of multi-layer $In_{0.4}Ga_{0.6}As$ QD samples grown under (a) As_4 flux and (b) As_2 flux, with nominal deposition of 8 MLs.

The use of an As_2 background to increase the QD separation along the chain can be understood by examining surface diffusion. As_2 molecules are more probable to find a favorable binding site before they desorb back into vapor phase because they do not have to break down on GaAs surface before bonding with Ga, unlike As_4. They fix Ga at stable site on surface faster than As_4 does, thus reducing the diffusion of Ga adatoms [12]. This leads to a decrease in the lateral intermixing of InGaAs along <0-11> during the dot formation. In Fact, As_2 background flux has been utilized to create an isotropic diffusion pattern that was necessary to for the growth of InAs quantum rings [13].

Strain Factor

Although in Figure 1(b) we have achieved uniform QDs that are well separated from each other, the 2D ordering is not isotropic yet. It seems that at indium composition of 40%, anisotropic strain transfer is still a dominant mechanism that allows dots to favor nucleation in separated chains. In fact, it has been shown that the strain in InGaAs over GaAs decreases rapidly only for indium composition less than 40% [14]. It is expected then that lower ndium composition would reduce the dependence on strain in ordering QDs, and at the same time diffusion will be the dominant mechanism. This will allow us to realize the advantage of using As$_2$ in achieving isotropic diffusion.

To confirm this assumption, we grew four multi-layer InGaAs sample at the same growth conditions under As$_2$ flux but with different indium composition and thus amount of deposition. Figure 2 shows the AFM images of samples with indium ratio of 0.3, 0.4, 0.5 and 1 (i.e. InAs) and nominal deposition of 15, 8, 6, and 2 MLs respectively.

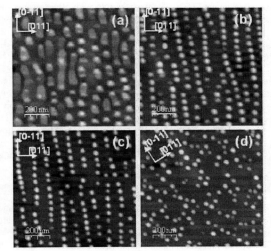

FIGURE 2. AFM top views of multi-layer InGaAs QD samples grown under As$_2$ flux with indium composition of (a) 0.3, (b) 0.4, (c) 0.4, and (d) 1, and nominal deposition of 15, 8, 6, and 2 MLs, respectively.

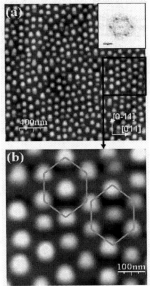

FIGURE 3. AFM top view of (a) multi-layer In$_{0.3}$Ga$_{0.4}$As QD sample grown under As$_2$ flux with nominal deposition of 13 MLs and (b) its magnification.

57

It is clear that higher indium composition results in a lower dot density due to the lower deposition and in short chain lengths and, eventually, no chains at all for the case of InAs. On the other hand, lower indium composition results in higher coverage that allows more time for diffusion to contribute to the ordering of QDs and uniformity of density distribution on surface as shown in Figure 2(a). It is noticed that the sample in figure 2(a) is the most promising for 2D ordering with the exception of the presence of coalescence between dots. Dots coalescence is understood to be due to increased coverage [15], so another sample with 30% indium composition was grown under As_2 flux with reduced coverage to 13 MLs. The AFM image of this sample is shown in figure 3 in two resolution levels.

By using As_2 as a background and 30% indium composition for vertically stacked layers of InGaAs QDs, we have achieved a good isotropic lateral ordering of dots as can be seen clearly in figure 3(b). This hexagonal geometry is also supported by the Fast Fourier Transform (FFT) of the AFM image shown in the inset of figure 3(a). This pattern can be explained in terms of the strain pattern that is transferred vertically. For example, we know that each QD transfers a vertical stain to the QD layer above. As stated earlier, when two dots are very close with in a layer the profile of the transferred strain pattern is over lapping. This results in only one nucleation site in the layer above from two sites below, and provides a mechanism to vertical order and thin out the QD density. As the density thins out we can see that each QD resides at nearly the center of a circle of radius R of influence in which there are no other QDs. As a result, assuming all QDs at the center of a circle of influence, six QD circles can neatly fit around one QD because there is no favored direction of diffusion anymore when using As_2.

CONCLUSION

In conclusion, we have shown experimentally that changing the growth environment from As_4 to As_2 breaks down the anisotropy of the ordered array of InGaAs QDs. We have attributed this change due the effect of As_2 molecules in changing the surface structure and the corresponding diffusion of adatoms over the GaAs <001> surface. QDs with very low indium composition were used to reduce the strain transmitted between QD layers and achieve better isotropic ordering and in fact a 3D ordered array of QDs.

ACKNOWLEDGMENTS

This work has been supported by National Science Foundation, Materials Research in Science and Engineering Center (MRSEC).

REFERENCES

1. D. Bimbreg, M. Gruandmann, and N. Ledentsov, *Quantum Dot Hetero-structure* (John Wiley & Sons, Chichester, 1999).
2. J. Phillips, *J. Appl. Phys.* **91**, 4590 (2002).
3. D. Leonard, M. Krishnamorthy, C. M. Reaves, S. P. Denbaars, and P.M. Petroff, *Appl. Phys. Lett.* **63**, 3203 (1993).
4. H. Lee, J. A. Johnson, J. S. Speck, and P. M. Petroff, *J. Vac. Sci. Technol. B* **18**, 2193 (2000).
5. Y. Nakamura, O. G. Schmidt, N .Y. Jin-Phillipp, S. Kiravittaya, C. Muller, K. Eberl, H. Grabeldinger, and H. Schwizer, *J. Crystal Growth* **242**, 339 (2002).
6. C. K. Hyon, S. C. Choi, S. H. Song, S. W. Hwang, M. H. Son, D. Ahn, Y. J. Park, and E. K. Kim, *Appl. Phys. Lett.* **77**, 2607 (2000).
7. Y. I. Mazur, W. Q. Ma, X. Wang, Z. M. Wang, G. J. Salamo, and M. Xiao, *Appl. Phys. Lett.* **83**, 987 (2003).
8. W. Q. Ma, M. L. Hussein, J. L. Shultz, and G. J. Salamo, *Phys. Rev. B* **69**, 233312 (2004).
9. Z. M. Wang, H. Churchil, C. E. George, and G. J. Salamo, *J. Appl. Phys.* **96**, 6908 (2004).
10. G. S. Solomon, *Appl. Phys. Lett.* **84**, 2073 (2004).
11. Z. M. Wang, S. Seydmohamadi, J. H. Lee, and G. J. Salamo, *Appl. Phys. Lett.* **85**, 5031 (2004).
12. C.G. Morgan, P. Kratzer, and M. Scheffler, *Phys. Rev. Lett.* **82**, 4886 (1999).
13. D. Granados and Jorge M García, *Appl. Phys. Lett.* **82**, 2401 (2003).
14. H. Li, *J. App. Phy.* **87**, 188 (2000).
15. G. S. Solomon, *Appl. Phys. Lett.* **66**, 991 (1994).

Growth of ZnO Nanorods from Zn and Zn- Zn$_3$N$_2$ Films

A. Toumiat[1,2], S. Zerkout[1,3], S. Achour[1], N. Tabet[4] and L. Guarbous[5]

[1] Constantine Ceramics Laboratory., Mentouri University, Constantine, 25000, Algeria
[2] Unité de développement de technologie de silicium, U.D.T.S. B.P 140 Alger-7 merveilles. 16200. Algeria
[3] Electro technical Department, University 20-08-55, Skikda, 21000. Algeria
[4] Surface Science Laboratory, King Fahd University of Petroleum and Minerals, Dhahran, Saudi Arabia
[5] Centre de Recherche Nucléaire, Alger, 16000. Algeria
toumiata@mail.usa.com

Abstract. This work reports on the growth of ZnO nanorods by simple thermal oxidation under air at 500°C of Zn and/or Zn-Zn$_3$N$_2$ precursors thin films. These films were deposited by reactive magnetron sputtering in argon atmosphere for Zn films and in a mixture of Ar – N$_2$ for Zn-Zn$_3$N$_2$. X- Rays diffraction (XRD) patterns of the precursors films showed the formation of both Zn and Zn$_3$N$_2$ phases before oxidation. However, after oxidation, X-ray Photoelectron Spectroscopy (XPS) analysis revealed the disappearance of Zn$_3$N$_2$ phase. This method allowed us to prepare high-density ZnO nanorods by oxidizing Zn-Zn$_3$N$_2$ precursor films. Scanning Electron Microscopy (SEM) investigations revealed that the films prepared from Zn-Zn$_3$N$_2$ showed higher density of nanorods and stronger photoluminescence with several mode phonon replicas at room temperature.

Keywords: ZnO nanorods, Zn$_3$N$_2$, Photoluminescence.
PACS: 78.55, 78.67-n, 81.07

INTRODUCTION

A considerable effort has been devoted to the synthesis and characterization of nanoscale materials because of their critical importance for the development of nanotechnologies. Their attractive physical properties enable them to play a major role in nanoscale devices. Nanostructured Zinc Oxide is one of the most important multifunctional materials that have been extensively studied during the last decade. It is a direct wide band gap semiconductor (3.37 eV) with a large exciton binding energy (60 meV) and many interesting optical, electrical and piezoelectric properties. The pioneering work of Kawasaki et al. [1] showed that room temperature excitonic laser emission was performed in ZnO nanocrystal. Since many authors [2-5] have been reported the lasing effect in ZnO nanostructures. Now, sun light has an increasing application in the production of clean energy via Photovoltaic (PV) solar cells. However, solar cells are currently expensive due to their fabrication cost and their low conversion efficiency performance. The use of nanomaterials has been suggested to enhance the performance of solar cells [6]. The growth of ZnO nanostructures have been successfully achieved by thermal oxidation of metallic Zinc [7-10]. They exhibit strong luminescence at room temperature. Recently, we have observed that when nitrogen is introduced during the sputtering of metallic zinc, we observe a higher density of nanorods after oxidation of the films [11]. These films showed an unusual photoluminescence dominated by many phonon replicas at room temperature.

EXPERIMENTAL

The synthesis of ZnO nanorods was conducted in two steps. The first was done through sputtering of metallic Zinc target, which was made from electrolytic zinc (purity 99.99 wt %) in Direct Current Magnetron sputtering machine

CP929, Nanotechnology and Its Applications, First Sharjah International Conference
edited by Y. I. Salamin, N. M. Hamdan, H. Al-Awadhi, N. M. Jisrawi, and N. Tabet
© 2007 American Institute of Physics 978-0-7354-0439-7/07/$23.00

(DC-Magnetron). Before deposition, the pressure in the chamber was 10^{-4} mbar. During the deposition the pressure was maintained at 10^{-2} mbar. Only Ar gas was used to prepare metallic zinc films, while a 70% Ar + 30% N_2 mixture was used to prepare $Zn-Zn_3N_2$ composites. The films were deposited on monocrystalline P-type Si(001) substrates. Second, the coated Si substrates were heat treated at 500°C in air. The oxidation was conducted for 5 hours. The samples were characterized using Grazing X-ray Diffraction (GXRD), Scanning Electron Microscopy (SEM), X-ray Photoelectron Spectroscopy (XPS) and Secondary Ion Mass Spectroscopy (SIMS) as well as Photoluminescence spectroscopy (PL).

RESULTS AND DISCUSSIONS

Figure 1 showed the XRD patterns of the as synthesized precursors. Six peaks corresponding to the hexagonal close packed crystal lattice appeared for the metallic zinc precursor films. When nitrogen is used as reactive gas, a mixed pattern of Zn and Zn_3N_2 peaks appeared.

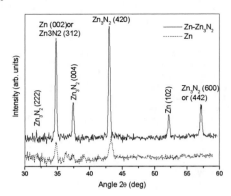

FIGURE 1. XRD patterns of thin films prepared by DC-magnetron: (solid) Zn_2N_3 30/70 % N2/Ar mixture, (dotted) Zn 100 % Ar.

The reported structure of Zinc Nitride [12-14] is a body cubic centered (bcc) with a cell parameter a = 0.977 nm. The measured d_{hkl} spacings, and cell parameters are given in Table 1.

TABLE 1. Phase determination of the Zn_3N_2 precursor.

Bragg angle	d(hkl) [Å]	Rel. Int. [%]	(hkl)	a [Å]
31.45	2.842	10	(222)	9.846
34.77	2.578	98	(312) or Zn (002)	9.646
37.41	2.402	53	(004)	9.608
42.93	2.105	100	(420)	9.414
52.17	1.752	40	Zn (102)	2.665
57.09	1.612	46	(600) or (442)	9.672

All peaks of Fig. 1 belong to the cubic structure except the peak at 2θ = 52.17°, which verifies the hexagonal plane spacing law, with a = 0.2665 nm and b = 0.4947 nm.

The cell parameter a varied from 0.933 nm to 0.984 nm, ehe average of 0.961 nm can be compared with the reported value of 0.977 nm [12].

The precursors were oxidized outside the chamber by heating the specimens for 5 hours at 500°C, in air. The ZnO obtained from metallic Zn films is called P1 and the ZnO obtained from Zn_3N_2-Zn composites is called P2. The XRD patterns for these two films, after oxidation, are shown in figure 2. Both films present polycrystalline crystal structure without preferential orientation and no diffraction patterns of other materials such as Zn and Zn_3N_2 were detected. This indicates that the zinc nitride formed during sputtering was completely transformed into ZnO. In fact, XPS and SIMS analysis did not detect any residual nitrogen in the films after oxidation. This allows us to conclude that all of the zinc nitride precursor has been transformed into zinc oxide.

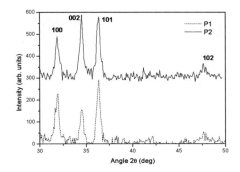

FIGURE 2. XRD patterns of ZnO thin films obtained by oxidation of Zn and Zn/Zn_3N_2 precursors at 500°C.

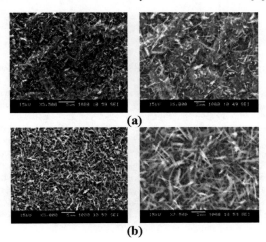

FIGURE 3. A general view of the ZnO nanorods: (a) without nitrogen in the precursor, and (b) with nitrogen in the precursor.

Figure 3 shows a general view of the obtained ZnO nanorods. The distribution of the rods is uniform throughout the Si substrates. In the case of P1, however, the nanorods are few, while in P2, much higher density of nanorods covering the surface was observed. These nanorods have an average length of 5 μm and an average diameter of 30 nm. They have a regular spatial and size distribution.

The role of zinc nitride on the growth of ZnO nanorods may be discussed on the basis of thermodynamic properties of this material. Wei et al. reported the growth of ZnO nano-crystal from an entire Zn_3N_2 precursor by hydrothermal process using NaOH and ethylene glycol solution [15]. They proposed a model where $[Zn (NH_3)_4]^{2+}$ act as precursor in the formation of nano-structural ZnO. In our case, the following reaction first can take place resulting in ZnO phase

$$Zn_3N_2(s)+\frac{5}{2}O_2(g)\rightarrow 2NO(g)+3ZnO(s),\qquad\qquad(1)$$

at the surface of the precursor with temperature increase. This reaction is very likely, since no trace of nitrogen was found after oxidation as it was verified by XPS and SIMS analysis (Fig. 4). In a second stage of nanorods growth, the oxygen diffused in the bulk and transformed Zn/Zn_3N_2 to ZnO with higher growth along [0001] direction. The differences in cell parameters between Zn_3N_2 and ZnO play an important role during the transformation and growth of ZnO nanorods. Dang et al.[16] have demonstrated the self catalytic growth of ZnO nanowires from metallic zinc by heating under oxygen atmosphere.

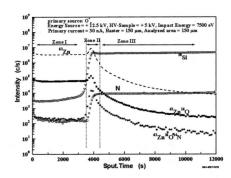

FIGURE 4. SIMS analysis, the nitrogen level in the ZnO layer in lower than in the Si substrate.

It is commonly admitted that PL spectra of ZnO nanostructures consist of two dominant bands; a high energy band associated with exciton recombination and a low energy band associated with structural defects such as oxygen vacancies [17]. In contrast to the PL spectra of bulk materials which can show a few phonon replicas, the PL spectra of nanostructures usually do not reveal such phonons. In this work, PL spectra were collected using Perkins Elmer LSB 50 spectrometer with Xenon (Xe) lamp as excitation source (λ= 325 nm). The results were plotted in Figure 5.

The PL intensity strongly increased in the case of P2. This may be due to the nanorods being abundant and well developed in P2 films. Moreover, a shoulder, approximately located at 3.30 to 3.31 eV, which is about 70 meV below the room temperature band gap of bulk ZnO (3.37 eV), can be distinguished in the spectrum of P1. Since the free exciton binding energy in ZnO is ~ 60 meV, this shoulder can be assigned to free exciton recombination. The maximum of the PL room temperature spectrum coincides with the emission of the 1LO phonon replica of the free exciton. It becomes the strongest feature, as was also found by [18]. This maximum is situated at about 3.24 eV, lower than the free exciton energy by the approximate amount of 60 meV. This value can be compared with $\hbar\omega-\Delta E\approx 60$ meV where $\Delta E\sim kT/2 = 12.5$ meV at room temperature and $\hbar\omega =72$ meV. The asymmetric shape of the PL spectrum with distinct shoulders and low energy tail can be attributed to the contribution from higher order phonon replicas to the emission. Six shoulders located at energy positions equal to 54.n meV, with respect to the free exciton energy, where n is an integer, are easily discernible down to 2.3 eV. They can be attributed to the non polar E_2 mode phono replicas, caused by vibration of oxygen atoms in ZnO, which matches the energy spacing between the observed peaks. The E_2 mode phonon replicas were, recently observed in the room temperature PL spectra of ZnO nanowires [19].

CONCLUSION

It was found that the presence of nitrogen in metallic zinc films considerably enhances nanorod growth during oxidation of the films at low temperature. The obtained ZnO films show room temperature photoluminescence dominated by numerous phonon replicas which are unusual in nanostructures.

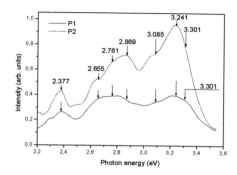

FIGURE 5. Room temperature PL spectra of ZnO thin films obtained by oxidation of Zn (P1) and Zn/Zn$_3$N$_2$ (P2) precursors at 500°C.

ACKNOWLEDGMENTS

The authors thank Dr. F. Abaidia from UMBB for GXRD measurements, Dr. K. Taibi from USTHB for SEM observations, Dr. M. Boumaour from UDTS for his valuable help and King Fahd University of Petroleum and Minerals for the XPS analysis.

REFERENCES

1. M. Kawasaki, A. Ohtomo, I. Ohkubo, H. koinuma. Materials Science and Engineering B **56** (1998) 239-245.
2. X. H. Zhang, S. J. Chua, A. M. Yong, H. D. Li, S. F. Yu and S. P. Lau. Appl. Phys. Lett. **88**, 191119 (2006).
3. K. Bando, T. Sawada, K. Asaka, Y. Masumoto, Journal of Luminescence **108** (2004) 385-388.
4. L. K. van Vugt, S. Ruhle and D. Vanmaekelbergh. Nanoletters **6**, 2707-2711 (2006).
5. J. C. Johnson, K. P. Knutsen, H. Yan, M. Law, Y. Zhang, P. Yang and R. J. Sakally, Nanoletters **4**, 197 (2004).
6. M. Law, L. E. Greene, J. C. Johnson, R. Saykally and P. Yang. Nature Materials **4**, 455 (2005).
7. Sunglae Cho, Jing Ma, Yunki Kim, Yi Sun, George K. L. Wong and John B. Ketterson, Appl. Phys. Lett. **75**, 2761 (1999).
8. Y. G. Wang, S. P. Lau, H. W. Lee, S. F. Yu, B. K. Tay, X. H. Zhang, and H. H. Hng. J. Appl. Phys. **94**, 354 (2003).
9. Y. G. Wang, S. P. Lau, X. H. Zhang, H. W. Lee, S. F. Yu, B. K. Tay, and H. H. Hng. Chem. Phys. Lett. **375**, 113 (2003).
10. H. J. Fan, R. Scholz, F. M. Kolb and M. Zacharias, Appl. Phys. Lett. **85**, 4142 (2004).
11. A. Toumiat, S. Achour, A. Harabi, N. Tabet, M. Boumaour and M. Maallemi, *Nanotechnology* **17**, 658 (2006).
12. K. Kurayama, M. Takahashi, and F. Suhahara. Phys. Rev. B **48**, 2781 (1993).
13. D. E. Partin, D. J. Williams, and M. O'keeffee. J. Solide State Chemistry **132**, 56 (1997)
14. F. Zong,a_ H. Ma, J. Ma, W. Du, X. Zhang, H. Xiao, F. Ji and C. Xue. Appl. Phys. Lett. **87**, 233104 (2005).
15. M. Wei, Z. Qi, M. Ichihara, I. Homma, and H. Zhou, Nanotechnology **18**, 095608 (2007).
16. H. Y. Dang *et al.*, *Nanotechnology* **14**, 738 (2003).
17. A. F. Kohan, G. Ceder, D. Morgan, and Chris G. Van de Walle. Phys. Rev. B **61**, 15019 (2000).
18. W. Shan, W. Walukiewicz, J. W. Ager III, K. M. Yu, H. B. Yuan, H. P. Xin, G. Cantwell, and J. J. Song, Appl. Phys. Lett. **86**, 191911 (2005).
19. S. Ramanathan, S. Bandyopadhyay, L. K. Hussey and M. Munoz, Appl. Phys. Lett. **89**, 143121 (2006).

CARBON NANOTUBES

Carbon Nanostructures: Synthesis and Characterisations

M. Bououdina

Department of Physics, College of Science, University of Bahrain, Kingdom of Bahrain, and
Advanced Materials Group, School M3, University of Nottingham, Nottingham, UK
mboudina@gmail.com

Abstract. Chemical vapor deposition technique has been used to produce a variety of carbon nanostructures. Type of catalyst (Fe, Ni) and its size, temperature, gas ratio C_2/H_4) and time affect the type of carbon produced as well as its length and diameter. Post-treatments (oxidation, reduction) have been carried out in order to remove the amorphous-capped carbon formed at the edges of tubes and fibers.

Keywords: Carbon, nanofiber, nanotube, CVD, XRD, TEM, XPS.
PACS: 81.05.Uw, 61.46.Fg, 81.65.-b, 68.37.Lp, 79.60.-i

INTRODUCTION

Carbon is a non-metallic element with the highest melting point at atmospheric pressure (triple point: 10 MPa, 4300-4700K) and a high affinity to bond with various atoms to form millions of compounds. It has a variety of forms starting from the hardest element, diamond, to the softest, graphite, including amorphous carbon, nanofoams, activated carbon, fullerenes, and carbon nanostructures (nanofibers and nanotubes [1]). Each type of carbon nanostructure contains various structural forms, which depend on the synthesis method, type of gas mixtures, type of catalyst and its size, and temperature. Carbon nanofibers and nanotubes have attracted great attention in the last decades due their high physical properties (electrical and thermal conductivity) and mechanical strength. Hence they present a wide range of potential applications: jewelry, electrodes for batteries, water purification, nanotechnology, electronics, optics, medical applications (poisoning, overdose, antibiotics for bacteria and cancer cells and tumor treatments), nanocoating, nanocomposites. This paper is about the synthesis and characterization (structure and microstructure) of carbon nanostructures.

EXPERIMENTS

Carbon nanostructures were prepared using either Fe_2O_3 or NiO powder as the catalyst precursor, by chemical vapor deposition (CVD). A typical experiment would use 0.05 g of catalyst and 2 hours reaction time. The temperature of the reaction was varied between 350 - 800°C. The feedstock gas was composed of ethene, C_2H_4 (99.9%) and hydrogen, H_2 (99.99%), using mass flow controllers to provide mixes of 100/0, 80/20, 50/50, 20/80 and 5/95, maintaining a total flow of 100 sccm. Samples were characterized using X-ray diffraction (Siemens D500 equipped with Kα-Cu λ=1.5418Å), transmission electron microscopy (JEOL JEM-2000FX II), high resolution TEM (JEOL JEM-4000FX). X-ray photoelectron spectroscopy (VG Scientific ESCA Lab x-ray photoelectron spectrometer fitted with a twin Mg/Al X-ray) and BET surface area analysis (Quantachrome Autosorb-1).

CP929, *Nanotechnology and Its Applications, First Sharjah International Conference*
edited by Y. I. Salamin, N. M. Hamdan, H. Al-Awadhi, N. M. Jisrawi, and N. Tabet
© 2007 American Institute of Physics 978-0-7354-0439-7/07/$23.00

RESULTS

Work reported in the literature investigating the effect of CVD processing conditions on the type of carbon nanostructures formed and the rate of carbon deposition tend to keep the reaction temperature the same (often 550-600°C). With the aim of identifying conditions that would yield high quantities of GNFs, the effect of reaction conditions on the carbon deposition over iron and nickel catalysts was investigated. In this paper, both temperature and gas ratio were changed using Fe and Ni catalysts in the form of oxides.

X-ray diffraction patterns reveal the formation of carbon nanostructure, with the precipitation of Ni or Ni_3C for Ni-catalyst (Figure 1a), and Fe_3C for Fe-catalyst (Figure 1b). The (002) peak of CNFs is very broad, which could be attributed to the presence of small crystallite size and also to some level of amorphisation within the fibres. The crystallite size was estimated from line broadening of (002) peak using Sherrer's formula. The crystallite size distribution seems to not vary with the conditions used hereafter; the average size is about 8 nm.

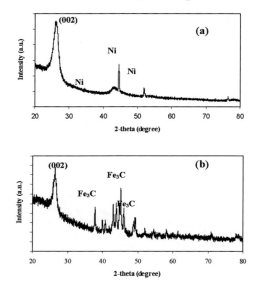

FIGURE 1. X-ray diffraction patterns.

The map below (Figure 2) illustrates that at low temperatures and for hydrogen concentrations $\geq 20\%$, herringbone nanofibres were formed. Increasing the temperature to 600°C and for the higher hydrogen concentrations (80 and 95%) yielded platelet nanofibres.

It can also be seen that at higher temperatures the predominant form of carbon was encapsulation (with carbon) of the catalyst particle with little or no filamentous carbon produced. However, although there was good control of the type of nanostructure formed, the deposition rates were rather low, with a maximum growth rate of 3.7 g $g_{cat}^{-1}h^{-1}$ at 500°C and a hydrogen concentration of 95%.

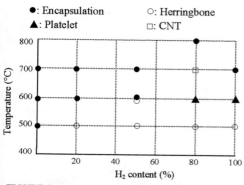

FIGURE 2. A process map showing the effect of temperature and feedstock composition on the type of carbon deposited over iron-based catalysts.

The nickel catalysts were found to be an order of magnitude more active than the iron catalysts. The yield is higher compared to Fe samples, ranging from 0.5 g $g_{cat}^{-1}h^{-1}$ for 350°C to 5.7 g $g_{cat}^{-1}h^{-1}$ for 500°C. It is interesting to note that the type of nanofibre formed changed from herringbone to platelet with increasing temperature, a similar trend to that found for the iron catalysts. However, the nickel catalysts suffered less with encapsulation and therefore had a wider processing window for nanofibre production.

TABLE 1. Some properties of the carbon nanostructures formed by CVD over nickel catalysts using an ethene to hydrogen ratio of 80/20 at temperatures from 350 to 700°C.

Sample	C-Nanostructure (catalyst phase)	Diameter from TEM (nm)	Surface area (m^2g^{-1})
GNFN700	Platelet (Ni)	300-20	49
GNFN600	Platelet (Ni)	500-20	67
GNFN500	Platelet (Ni)	500-20	170
GNFN400	Herringbone (Ni₃C)	200-20	138
GNFN350	Herringbone (Ni₃C)	200-20	252

For Fe-catalyst, the surface area slightly changes with temperature, ranging from 45 – 70 m².g⁻¹. However, for Ni-catalyst, the surface area increases with decreasing temperature, from 50 m².g⁻¹ for 700°C) up to 250 m².g⁻¹ for 350°C.

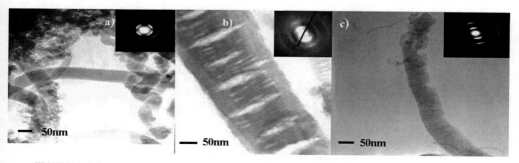

FIGURE 3. Effect of temperature and feedstock composition on the type of carbon deposited over Fe-based catalysts.

Figure 3 gives TEM images and selected area diffractions (SAD) for three samples. GNFN400 was a mix of regular herringbone GNFs with a diameter ca. 70 nm and much smaller helical GNFs with diameters down to 20 nm (Figure 3a). GNFN500 produced very wide platelet structures, but as can be seen from the broad arcs in the SAD had less regular graphene layers (Figure 3b). GNFN700 produced much more regular platelet GNFs (Figure 3c). An important result coming out of the low temperature CVDs was that the catalyst phase after reaction was nickel carbide Ni_3C, rather than metallic Ni. Herringbone GNFs were only produced on nickel when the catalyst phase was the carbide, indicating preferential GNF formation depending on the catalyst phase.

An initial investigation of tubes grown for a reaction time of 2 hours found that those produced at 600°C were more abundant, longer and had larger external diameters than those produced at 700°C for which more carbon encapsulation of the nickel occurred. Tube dimensions were of the order of 1-5 μm in length and 20-100 nm in external diameter at 600°C, and 1-2μm in length and 15-70nm in external diameter at 700°C, respectively.

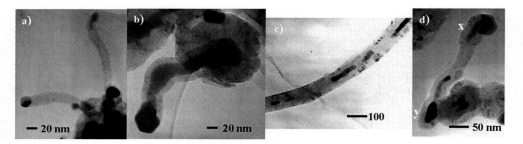

FIGURE 4. (a,b) nanotubes grown at 700°C for 10 and 20 minutes, respectively. (c) Nickel catalyst trapped in the hollow core of a nanotube, indicating a highly mobile state has been attained (700°C, 2 hrs). (d) Nanotube with faceted (x) and encapsulated (y) Ni particles at opposite ends (700°C, 20 min).

Figures 4a,b illustrate MWNTs formed following 10 and 20 minutes of growth at 700°C, respectively. The tubes for these shorter reaction times were typically 100 to 500nm in length, and often found emanating from larger nickel particles encapsulated in carbon. Regardless of reaction time and temperature, highly elongated Ni particles were often situated at both ends of each tube and on occasion distributed along their hollow cores (Figure 4c), the morphology of which might indicate that the catalyst had attained a highly mobile or molten state during synthesis, despite the melting point of Ni being 1453°C.

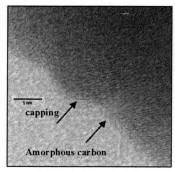

FIGURE 5. HR-TEM image showing capping and amorphous carbon at the edge of a platelet GNF.

Upon closer inspection of the GNFs using HR-TEM it became evident that over the whole of the surface of the GNFs was a layer of less-ordered carbon (see Figure 5). This carbon could take a number of forms: amorphous carbon sheath; caps connecting graphene edges; and realignment of the graphene layers becoming parallel with the fibre axis toward the edges of the GNF. This less-ordered carbon will obviously act as a barrier to hydrogen diffusing in between the graphene layers and methods of removing it and exposing the graphene edges were investigated.

In order to remove the less-ordered carbon present, a selective oxidation treatment using CO_2 gas was undertaken. DTA-TG was used to identify the onset of oxidation, which was 600°C for GNFN700 and GNFN500. Oxidation treatments were carried out in a tube furnace, at the onset temperature, under flowing CO_2 to investigate the rate of weight loss. As can be seen from Table 2, a linear oxidation rate was measured under these conditions and for reaction times up to 3 hours (resulting in a weight loss of 6 wt%).

TABLE 2. Weight loss of GNF3 samples after CO_2 treatments.

Time (min)	Weight loss (%)
15	0
60	2.7
180	6.1

XPS analysis shows that the O_2 peak is more pronounced after CO_2-treatment, about 3 times of O_2 content has been observed, as show in survey specters Figure 6.

The oxygenated species on the GNF surface after CO_2 selective oxidation were removed by reacting with hydrogen at 400°C. Reduction of a CO_2-treated GNFN3 sample under Ar – 5%H_2 gas (0.2 bars) at 400°C has been carried out for 2 hours in a furnace. The weight loss has been estimated to be around 1.5 wt.%.

Observation of the graphitic planes using HREM around the catalysts and along the tube walls confirmed that the nanotubes displayed varying levels of crystallinity. As a general observation the carbon walls followed the shape of the catalyst but then transformed with increasing distance from the catalyst into the parallel-sided walls of the NTs, whilst also become more disrupted and broken up.

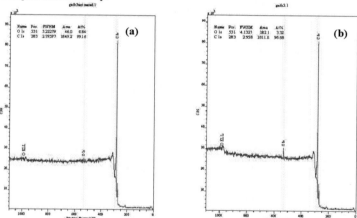

FIGURE 6. XPS of (a) untreated GNFN3 and (b) CO_2-treated GNFN3 samples.

Mechanism of Formation of Nanotubes

Starting from the initial transformation of NiO to Ni, these combined observations implicate the following mechanism in the formation of MWNTs trailing behind clean faceted Ni particles (Figure 7):

71

- On the surface of a catalyst particle, ethene decomposes to carbon that is adsorbed, whilst hydrogen is released. A resultant carbon monolayer presumably covers the exposed faces of the particle (Figures 7i,ii).
- It is suggested that the Ni faces that have a high carbon diffusion rate preferentially absorb the carbon atoms, creating a low C-content Ni-alloy. (The combined TEM and XRD evidence confirmed that distinct nickel carbide had not formed.)
- A concentration gradient of carbon will presumably develop within the particle between the catalyst-vapor surface (Figure 7ii(x)) and the Ni-support interface (Figure 7ii(y)). The absorbed carbon will thus undergo bulk diffusion through the Ni to the precipitating planes where individual atomic planes of carbon are produced and sequentially build up parallel to the trailing surface of the Ni particle.
- This precipitation of carbon causes the propulsion of the Ni particle away from its original location (Figure 7iii).
- As the catalytic reaction proceeds more carbon is adsorbed, absorbed and diffused through the Ni particle to be precipitated and perpetuate the growth of the nanotube (Figure 7iv). During this process the nickel apparently approaches an almost liquid state.

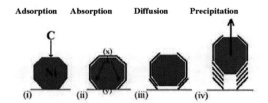

FIGURE 7. Proposed nanotube growth mechanism.

CONCLUSION

A variety of carbon (nanotubes, nonofibers) has been synthesized via CVD technique using C_2H_4/H_2 mixture and NiO/Fe_2O_3 catalysts. Amorphous and capped carbons formed at the edges have been removed by subsequent CO_2/H_2 treatment.

REFERENCES

1. S. Ijima, *Nature* **354**, 56, (1991).

Factors Affecting Carbon Nanotubes (CNTs) Synthesis via the Chemical Vapour Deposition (CVD) Method

M. Zakaria and S. M. Shariff

PETRONAS Group Research, Lot 3288&3289, Off Jalan Ayer Itam, 43000 Kajang, Selangor, Malaysia
muzdalifah@petronas.com.my

Abstract. Several studies have shown that the presence and type/size of catalysts affect the growth mechanism and the nanotubes structures. The catalysts are believed to affect the catalytic decomposition of the carbon feedstock into carbon atoms and subsequent diffusion of these atoms through the catalyst particles. Numerous studies have also demonstrated that saturated hydrocarbon gases such as methane and CO tend to produce fewer nanotube walls and hence are a more favourable feedstock for single wall nanotube growth. Likewise, unsaturated gases with higher carbon content such as acetylene and benzene are typically used to grow multi wall carbon nanotubes. Other factors such as reaction time and temperature, inert gas source and gas flow rates are also known to affect the synthesis process. Major trends will be outlined to show the correlation between some of these different parameters with respect to the synthesis process and final products yield in order to achieve the optimum processing condition for CNTs.

Keywords: carbon nanotube, catalytic CVD, carbon feedstock, catalyst.
PACS: 81.07.De

INTRODUCTION

Carbon nanotubes (CNTs) are one of the most widely studied materials in nanoscience and nanotechnology R & D. This is particularly due to the amazing properties that these materials display and their potential applications in various industries. Since its discovery in 1990, CNTs have grown from lab-experimented material to a commercially available product. Advancement in the processing methods and improvement in the product properties have led to current application of the nanotubes in sensing devices, flat panel display and other field emission devices. Future anticipated applications include, among others, hydrogen storage material, supercapacitors, polymers and composites additives and advanced batteries electrodes.

One of the most favored methods in producing CNTs is via the chemical vapor deposition (CVD) technique which generally involves decomposing a hydrocarbon gas, such as methane, into carbon atoms which then condense on a cooler substrate normally coated with various catalysts to facilitate the tubes growth. This method has several advantages over arc discharge and laser ablation techniques in better processing control, lower reaction temperature (between 550-1000°C) and potential for production scale-up. However, it is worth to note that the current CNTs application is still limited by the amount of yield and relatively complex production techniques which tend to drive the cost up. Hence ongoing works are focused on finding ways to improve the product yield through process scale-up to achieve mass production and at the same time reduce the complexity of the production technique and processing control.

Even though the determination of the controlling parameters in CNTs production via catalytic CVD is widely studied, little is still known about the actual effect of carbon feedstock, catalyst types and sizes, catalyst/substrate interaction, growth temperature, reaction time and gas flow rates on the nanotubes physical and mechanical properties,

CP929, *Nanotechnology and Its Applications, First Sharjah International Conference*
edited by Y. I. Salamin, N. M. Hamdan, H. Al-Awadhi, N. M. Jisrawi, and N. Tabet
© 2007 American Institute of Physics 978-0-7354-0439-7/07/$23.00

crystallinity, purity and product yield. This paper will attempt to review the effect that these different parameters have on CNTs production and draw major correlations between some of these parameters.

GROWTH MECHANISMS

A significant body of research work has been done on trying to understand the mechanisms of nanotube growth where several growth models have been put forward. It is generally believed that CNTs growth via catalytic CVD involves the diffusion and precipitation process, i.e., carbon containing gas decomposes into carbon atoms at high temperature, these atoms will in turn diffuse into the catalyst either by surface diffusion or bulk diffusion. When supersaturation is reached, carbon will precipitate out to form nanofibers. The main roles of the catalyst are to lower the activation energy of the chemical reaction, thus increasing the reaction rate, and to provide nucleation sites for nanotube growth. In addition, several catalysts are known to react with the carbon atoms forming surface carbides. Since these carbide particles are unstable, they will further decompose, releasing carbon atoms through the catalyst.

Chhowalla et al. [1] described the tip growth mechanism for their work on the growth of vertically aligned carbon nanotubes on supported catalysts. In this model carbonaceous gas (acetylene) decomposes on the surface of the catalyst particles (Ni), the carbon diffuses through the particles and precipitates out on the opposite side leaving the catalyst particle to remain at the tip of the growing tube. The rate limiting step in this multi-walled tube growth is thought to be the carbon diffusion through the catalyst [2] which infers the strong influence of catalyst size on the growth process. On the other hand, carbon supply to the catalyst is believed to be the rate limiting step for SWNT growth hence influence of partial pressure of the gasses becomes significant. A variation to this tip-growth model is the base-growth mechanism where a stronger catalyst-substrate interaction exists and carbon diffuses from the sides into the catalyst particles resulting in growth from the bottom up and the catalyst particles remain at the root of the growing tubes.

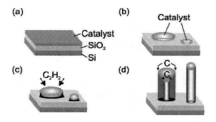

FIGURE 1. Tip growth model as proposed by Chhowalla et al. [1]. (a) Si substrate with a thin SiO_2 barrier layer and the Ni catalyst layer, (b) fragmentation of catalyst into nanoparticles by sintering, (c) decomposition of C_2H_2 on the top surface of the nanoparticle, (d) growth of nanotubes below Ni catalyst by carbon diffusion through the Ni particle

CARBON FEEDSTOCK

It is commonly observed that carbon feedstock such as methane (CH_4) [3-6] and CO [7-8] will normally yield SWNTs while acetylene (C_2H_2) [9-14] and ethylene (C_2H_4) [15] will produce MWNTs. Nevertheless, several studies have reported that under certain conditions these may not always be the case. A study carried out by Hafner et al. [2], successfully grown SWNTs ($\varnothing$ = 0.5-3 nm) from ethylene on Fe/Mo alumina supported catalyst by limiting the partial pressure of ethylene. This resulted in lower carbon supply and reduced the carbon concentration in the catalyst particles which allowed the tubes to form more slowly. Similar study carried out by Paillet et al. [16] produced SWNTs ($\varnothing$ = ~1.5 nm) from ethylene on Ni nanoparticles catalyst where improvement on synthesis yield was achieved also by lowering the ethylene flow. Both these studies signify the strong correlation between the rate-limiting step of carbon supply to the catalyst and SWNTs growth mechanism. Likewise, work by Liao et al. [8] did not produce any SWNTs using CO which was attributed to the large catalyst particles size (~10 nm) which affect the wall formation. Temperatures and carbon supply were also seen to have negligible effects on the CNTs morphologies. Another similar study [17] produced straight, onion- shaped and herringbone nanofibres by using methane as the carbon source. These results seem to suggest that by controlled manipulation of processing parameters such as catalyst size, reaction temperatures and time and gas feed rate the desired type of CNTs (either single-walled or multi-walled) can be produced regardless of the type of carbon source used. Growth of SWNTs also required a relatively higher temperature compared to MWNTs since a higher reaction temperature will permit the formation of defect-free crystalline structures. For this reason, methane is a suitable feedstock since it is known to be kinetically stable against pyrolytic decomposition at high temperatures.

Less conventional carbon feedstock has also been used in several studies. Qiu et al. [18] used coal as the carbon source and observed formation of SWNT bundles with average diameter of 1-2 nm. The CH_4, CO and H_2 components in the coal were believed to take part in the formation of the CNTs. Xu et al. [19] used CO_2 over Fe/CaO catalyst to form MWNTs with average diameter of 50 nm. Reaction temperatures and catalyst support materials were found to have a strong effect on the growth processes where CNTs will only formed between temperature range of 790 to 810°C. Likewise, typical catalyst supports such as alumina and silica were found to inhibit the growth process and no product was yielded. Thiruvengadachari [20] used cyclohexane (C_6H_{12}) and Cu-Ni as the carbon source and catalyst respectively to produce MWNTs. Strong correlation was observed between catalyst diameter and CNTs diameter in this work where CNTs with average diameter of 50-60 nm grew on catalyst particles size of about 60 nm.

CATALYST

Transition metals, Fe, Co, Ni and their combinations are the most common catalysts used in CNTs production. Catalysts are known to affect the nanotube's growth mechanisms and structures. For bimetallic systems, i.e., combination of two metal catalysts, it was found that composition and distribution of each metal in the catalyst particles affect the nanotube morphologies. Liao et al. [8] showed that higher Mo content in Co-Mo catalyst led to lower catalyst activity and reduced the decomposition rate of the carbon feedstock and hence decreased the production rate of carbon atoms. This allowed all the carbon atoms to diffuse through the catalyst for continuous CNT growth.

The amount of catalysts can also affect the type of nanotubes produced. Tang et al. [5] showed that addition of Mo to Co/MgO catalyst can increase SWNTs yield up to a certain limit where above this MWNTs were formed. This was further supported by Yoon et al. [11] where addition of Mo to Co-Mo catalysts was seen to increase the catalytic effect in producing SWNTs and by Nagaraju et al. [12] where mixture of Fe and Co resulted in the formation of MWNTs at 600°C where none was observed when only Fe or Co were used as catalysts.

Synthesis temperature and pressure also have major effects on the structure of nanotubes and product yield. Li et al. [21] showed that at low gas pressure of 0.6 Torr the structure changed from hollow to bamboo-like with an increase in temperature from 600-1050°C and maximum yield of 400% was achieved at 800°C after which the yield started to decrease. In contrast, at higher pressure of 760 Torr no change in morphology occurred to the bamboo-like structures that initially formed and maximum yield of 700% was achieved at higher temperature of 900°C after which the yield also started to decrease. The decrease in product yield was attributed to the reaction between iron catalyst with carbon forming stable iron carbide which occurred at high temperature and thus reducing the catalytic activity for CNTs growth. On the contrary, Fonseca et al. [9] did not observed any nanotube formation at 750 and 800°C by employing the same carbon source and catalyst and this may likely be due to the different catalyst preparation method employed, the smaller amount of catalyst used and shorter reaction time. It is also generally observed that an increase in temperature leads to lower density of tubes due to increase formation of amorphous carbon. On the other hand, high growth temperature may be necessary to obtain defect-free tubes as any defect will anneal out to produce well-crystalized CNTs. Careful control during high temperature synthesis is therefore required so that a balance between high crystallinity and high purity products can be achieved.

Catalyst support affects the synthesis of CNTs to a lesser extent as shown by various studies in which less conventional types of substrates have been employed such as vanadium plates [22] and graphite fibers [23] where tubes formation was still observed. Nevertheless, Xu and Huang [19] did not observe any nanotube growth using CO_2 on Al_2O_3, SiO_2 and MgO substrate. Only by using CaO support CNTs were able to be produced. More studies are required in this area to elucidate the relationship between catalysts support and carbon feedstock.

Several attempts to grow nanotubes without a catalyst, i.e., directly on the substrate have also been made. A successful result was obtained by Du and Pan [15] where nanotubes were grown on Ni substrate at temperatures ranging from 650-850°C. Temperature was observed to play an important part where no tubes could be detected at 850°C and this was attributed to the effect temperature has on the nucleation and growth mechanism.

SUMMARY

It has been shown by numerous studies that several parameters, namely; carbon source, catalyst type and size, temperature, pressure and reaction time, play significant roles in CNTs growth via catalytic CVD method. Nonetheless, it is difficult to conclusively pin-point the individual effect these parameters have on the growth process as they are closely interrelated and often dependent on each other. Hence, no single dominant parameter acts as the driving force for the whole process. The effect of carbon source is normally coupled with reaction temperature and pressure due to gas stability and the activation energy effect. Different metal catalysts were shown to interact differently when exposed to the same carbon feedstock and reach their optimum catalytic activities at different temperatures, hence no clear relationship is evident. However, in general it is believed that catalyst type affects the CNTs morphologies and the tube diameter is dependent on the catalyst particle size. Catalyst effect is seen to have a larger impact on the growth process and CNTs structures.

REFERENCES

1. M. Chhowalla et al., *Journal of Applied Physics* **90**, 5308-5317 (2001).
2. J. H. Hafner et al., *Chemical Physics Letters* **296**, 195-202 (1998).
3. J. Kong, M.H. Cassell and H. Dai, *Chemical Physics Letters* **292**, 567-574 (1998).
4. J.F. Colomer et al., *Chemical Physics Letters* **317**, 83-89 (2000).
5. S. Tang et al., *Chemical Physics Letters* **350**, 19-26 (2001).
6. Z. Li et al., *Carbon* **40**, 409-415 (2002).
7. B. Kitiyanan et al., *Chemical Physics Letters* **317**, 497-503 (2000).
8. X. Z. Liao et al., *Applied Physics Letters* **82**, 2694-2696 (2003).
9. A. Fonseca et al., *Applied Physics A* **67**, 11-22 (1998).
10. C. J. Lee, J. Park and J.A. Yu., *Chemical Physics Letters* **360**, 250-255 (2002).
11. Y. J. Yoon et al., *Chemical Physics Letters* **366**, 109-114 (2002).
12. N. Nagaraju et al., *Journal of Molecular Catalysis A: Chemical* **181**, 57-62 (2002).
13. K Hernadi et al., *Materials Chemistry and Physics* 77, 536-541 (2002).
14. Y. Yao et al., *Journal of Materials Science: Materials in Electronics* **15**, 583-594 (2004).
15. C. Du and N. Pan, *Materials Letters* **59**, 1678-1682 (2005).
16. M. Paillet et al., *Diamond & Related Materials* **14**, 1426-1431 (2005).
17. N. Zhao et al., *Mater. Sci. Eng. A (2007)*, doi:10.1016/j.msea.2007.01.051.
18. J. Qiu et al., *Fuel Processing Technology* **85**, 913-920 (2004).
19. X. Xu and S. Huang, *Materials Letters (2007)*, doi:10.1016/j.matlett.2007.01.059.
20. B. Thiruvengadachari and P.K. Ajmera, *Mater.Lett. (2007)*,doi:10.1016/j.matlett.2007.01.108.
21. W. Z. Li, J. G. Wen and Z. F. Ren, *Applied Physics A* **74**, 397-402 (2002).
22. J. W. Seo et al., *Applied Catalysis A: General* **260**, 87-91 (2004).
23. S. Zhu et al., *Diamond & Related Materials* **12**, 1825-1828 (2003).

Application of Taguchi Method in the Optimization of ARC-Carbon Nanotube Fabrication

M. Jahanshahi[1], R. Jahan-Bakhsh[2], H. Solmaz[1] and J. S. Razieh[1]

[1] *Nanobiotechnology Research Lab., Faculty of Chemical Engineering, Babol University of Technology, Babol, Iran*
mmohse@yahoo.com
[2] *Electroanalytical Chemistry Research Lab., Department of Chemistry, Faculty of Basic Science, Mazandaran University, Babolsar, Iran*

Abstract. Carbon nanotube, a new form of element carbon, is composed of graphene sheets rolled into closed concentric cylinders with diameter of the order of nanometers and length of micrometers. Since its discovery in 1991, extensive applications have been found in physical, chemical and material science fields. Due to its novel electronic structure, in electrochemical area, many researches have been carried out to study its electrocatalytic behavior towards many substances such as O_2, H_2O_2 and NADH utilizing its ability of fast electron transfer. Particularly, carbon nanotube's biocompatibility together with its catalytic ability to H_2O_2 and NADH opened its application in fabricating excellent amperometric nano(bio)sensors. Arc-in-liquid method has been developed to synthesize many kinds of nano-carbon structures such as carbon onions, carbon nanohorns and carbon nanotubes. This technique is considered as a low cost method because it does not require any expensive equipment. In here, this method was used for fabrication of nanotubes, and then a modified acid treatment method applied for purification stage. To optimize the synthesis, four various factors (voltage, current, catalyst and plasma) which are mostly affected on the carbon nanotubes were introduced to the Taguchi software in four levels. This statistical experimental design can determine the effect of the factors on characteristic properties and the optimal conditions of factors.

Keywords: Carbon nanotube, liquid arc discharge, synthesis, optimization, Taguchi method.
PACS: 81.07.De

INTRODUCTION

The number of publications on carbon nanotubes (CNTs) and CNT-reinforced composite materials has grown very quickly since CNTs were discovered by Iijima [1] more than a decade ago, as evidenced by several recent review articles on mechanical behavior, fabrication and applications of such materials [2].

Three typical synthesis methods have been developed for the production of CNTs, including the conventional arc discharge in an inert gas (argon, helium) atmosphere (or hydrogen), laser vaporization and chemical vapor deposition (CVD) [3]. Traditional arc discharge requires a complicated vacuum and heat exchange system. Laser vaporization produces high quality CNTs but demands considerable power. The yields of the above two methods are very low (~mg/h) [4]. In the other hand, it is also hard to get high-quality CNTs through CVD technique yet [3]. Arc discharge in liquid environments is a new method of synthesizing CNTs developed recently. All that is required is a dc power supply and an open vessel full of liquid nitrogen, deionized water or aqueous solution. This method does not require vacuum equipment, reacted gases, a high temperature furnace and a heat exchange system. Indeed the liquid substituted both vacuum and cooling system [5]. Consequently, this method is extremely simple and cheap [4]. Until now different types of carbon nanostructured have been successfully synthesized by this method [3]. Properties of carbon nanotube obtained

CP929, *Nanotechnology and Its Applications, First Sharjah International Conference*
edited by Y. I. Salamin, N. M. Hamdan, H. Al-Awadhi, N. M. Jisrawi, and N. Tabet
© 2007 American Institute of Physics 978-0-7354-0439-7/07/$23.00

by this method are affected by various parameters such as the voltage, current, types of the catalyst, types of solution and so on. The interrelationships between the above parameters are complex, and the analysis of this system to optimize the factors is a time consuming work. Hence, the analyses using conventional experimental methods are inefficient. The efficient analyses of the complex system using the Taguchi method has been performed recently [6]. The objectives of this work are: (a) to evaluate the effect of several parameters on its synthesis, and (b) to apply Taguchi method on the optimization of properties and to obtain the acceptable carbon nanotube in different ways by using optimal synthesis conditions.

MATERIALS AND METHODS

Materials

All chemicals were of analytical grade and aqueous solutions were prepared with doubly distilled water. The arc discharge apparatus employed in this study comprised an open vessel, graphite electrodes, and a DC power supply. Diameters of cathode and anode were 6 mm and 12 mm, respectively. The anode was drilled a 4 mm diameter hole and filled with mixture of catalysts powder in each experiment which was needed.

Fabrication and Purification Process

The brass electrode holders were free to move forward and backward, which enabled proper electrode gap adjustment to be made during arc discharge (Figure 1). It should be noted that the discharge in liquid environments is erratic, thus it is critical to control precisely the arc gap in order to run the arc continuously. Upon ignition, to avoid arc disruption, the electrode gap was maintained 1 mm. We used a digital controllable power supply with constant voltage output up to 120 V. The arc lasted for 120 seconds. The discharge current for all experiments sustained at 100A at which the arc showed to be quite stable and the voltage was adjusted for each experiments [7]. Evaporation of the solution during the arc discharge is considered to be negligible. Different metallic catalysts with various ratio were applied which are going to explain. Taguchi's orthogonal array table was used to optimum the weight of the pure carbon nanotubes by choosing four parameters that could affect the synthesis of carbon nanotube. Table 1 shows the parameters and levels used in this experiment.

The as-prepared deposit was purified using different combinations of the following purification steps for all tests: first, sonicating 0.5 g of as-prepared deposit in 12 N HCl for 30 minutes and leaving the resulting solution to stand overnight. Then it was refluxed in 6 N HCl for 6 hours. After treating the deposit with acid, the resulting solution is diluted with distilled water and centrifuged. After centrifugation, the supernatant solution is decanted and the residue is transferred onto a Millipoor filter paper. The residue on the filter paper is washed with distilled water (4–5 times) to remove the acid. The residue is dried in an oven at 80–100 °C for 5 hours. Then the carbon nanotubes were weighted and kept in desiccators [8, 9].

RESULTS AND DISCOSSION

SEM Images of Synthesized Carbon Nanotubes

Figure 2 illustrated the SEM images of carbon nanotube before and after purification. Figure 2b shows the SEM image of purified carbon nanotubes in which some impurities have been removed by the describe method. By all appearances it is cleaner than that in Figure 2a, which is the SEM image of as-grown CNTs. Contrasting the two images we can conclude that the carbon and catalyst impurities are almost completely removed in this process. The other SEM images and Raman spectroscopy proofed this claim too (results not shown).

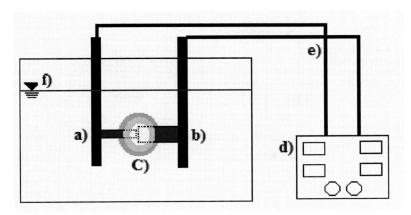

FIGURE 1. The schematic setup arrangement,a) cathode graphite b) Anode graphite c) Arc Plasma d) DC Arc Supplier e) wires f) Water surface

TABLE 1. Parameters and levels used in this experiment.

Factors	Levels			
A: Solution	Dionized water	NaCl (0.25 M)	KCl (0.25 M)	LiCl (0.25 M)
B: Voltage	35	30	25	20
C: Catalyst	Ni/Ni	Ni/Fe	Ni/Mo	Ni/Co
D: Catalyst ratio	1/1	1/2	1/3	1/4

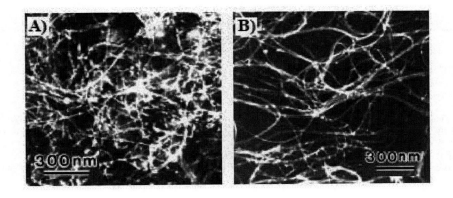

FIGURE 2. The SEM images of CNTs A) before B) after purification stage

Taguchi Array Design

A Taguchi orthogonal array design was used to identify the optimal conditions on the synthesis of carbon nanotube Duo to the identification between single or multi wall carbon nanotubes was too difficult; we consider optimizing th net weight of carbon nanotubes after purification. Table 2 shows the structure of Taguchi's orthogonal array design an the results.

TABLE 2. Experimental measured values for carbon nanotube weight and S/N ratio (Taguchi orthogonal array table L-16)

Experiment	Experimental conditions				Net weight (mgr)		S/N ratio (dB)
	A	B	C	D	Run 1	Run 2	
1	1	1	1	1	1.2	2.8	3.861
2	1	2	2	2	2.4	3.1	8.535
3	1	3	3	3	4.1	4.6	12.726
4	1	4	4	4	5.5	5.1	14.466
5	2	1	2	3	1.9	2.5	6.605
6	2	2	1	4	3.2	3.0	7.989
7	2	3	4	1	5.3	5.5	14.643
8	2	4	3	2	4.2	3.7	11.879
9	3	1	3	4	1.5	2.3	4.992
10	3	2	4	3	3.7	3.2	10.687
11	3	3	1	2	4.6	4.1	12.726
12	3	4	2	1	4.2	3.9	12.131
13	4	1	4	2	5.6	5.1	14.538
14	4	2	3	1	2.9	3.5	9.988
15	4	3	2	4	3.6	4.1	11.654
16	4	4	1	3	3.8	3.1	10.622

Determination of Optimal Conditions

Table 3 shows the main effects and it can be understand from that the voltage and type of catalyst have the mo effects on the carbon nanotube synthesis. The optimum conditions can be determined through the response table of th ANOVA-TM software. Therefore, based on the S/N and ANOVA analyses, the optimal parameters for carbon nanotub production are the solution LiCL (level 4), 25 V as voltage (level 3), Ni/Mo as catalyst (level 4) and the catalyst rati must be ½ (level 2). The estimate result was 6.52 mgr pure carbon nanotubes were achieved in optimum conditions. B doing the optimum test, 6.89 mgr cleaned carbon nanotubes were synthesis.

CONCLUSION

In summary, we used a simple DC arc-discharge method in liquid media with easy handling equipment for synthesi the carbon nanotubes. The method could be scaled up and used for continuous production [10]. In this study, Taguch design method was used to optimize the parameter values for obtaining desired characteristics. Voltage and catalyst ar the parameters having major effects on the fabrication of CNTs. By optimal conditions (at 25v voltage, LiCl as a salt in solution, Ni/Mo as a catalyst and the ½ as a catalyst ratio) in this method, 6.89 mgr pured CNTs were produced and this result has a good agreement with data analyzed by Taguchi method. Applying the synthesis CNTs on GC electrodes as nanobiosensors is going to be our next research project.

TABLE 3. The ANOVA table of net weight of carbon nanotube synthesis

	Factors	Degrees of freedom	Sum of Squares	Variance	F-Ratio
1	Solution	3	7.899	2.63	0.548
2	Voltage	3	78.043	26.014	5.422*
3	Catalyst	3	52.364	17.788	3.707*
4	Catalyst ratio	3	11.209	3.736	0.778
	error	3	14.392		
	Total	15			

* Main significant parameter

ACKNOWLEDGMENT

The Authors would like to thank Nanobiotechnology Research Laboratory. Also special thanks to Babol University of Technology and Electroanalytical Chemistry Research Laboratory of Mazandaran University for supporting this research.

REFERENCES

1. Iijima S. *Nature* **354**, 56–8 (1991).
2. Z. Xu, X. Chen, X. Qu, J. Jia, and Sh. Dong, *Biosensors and Bioelectronics* **20**, 579–584 (2004).
3. H. W. Zhu, X. S. Li, B. Jiang, C. L. Xu, Y.F. Zhu, D.H. Wu, and X. H. Chen, *Chemical Physics Letters* **366**, 664–669 (2002).
4. Sh. Wang, M. Chang, K. M. Lan, Ch. Wu, J. Cheng, and H. Chang, *Letters to the Editor / Carbon* **43**, 1778–1814 (2005).
5. H. Lange, M. Sioda, A. Huczko, Y. Q. Zhu, and H. W. Kroto, *Carbon* **41**, 1617–1623 (2003).
6. J. Ting, Ch. Chang, Sh. Chen, D. Lu, Ch. Kung, and F. Huang, *Thin Solid Films* **496**, 299 – 305 (2006).
7. H. Huang, H. Kajiura, A. Yamada, and M. Ata, *Chemical Physics Letters* **356**, 567–572 (2002).
8. R. B. Mathur, S. Seth, Ch. Lal, R. Rao, B. P. Singh, T. L. Dhami, and A. M. Rao, *Carbon* (in press).
9. Y. Feng, G. Zhou, G. Wang, M. Qu, and Z. Yu, *Chemical Physics Letters* **375**, 645–648 (2003).
10. N. Sano, H. Wang, M. Chhowalla, I. Alexandrou, and G. A. J. Amaratunga, *Nature (London)* **414**, 506 (2001).

Finite Element Model of SWCNT Under Hydrostatic Pressure

A. Sakhaee-Pour and M. T. Ahmadian

*[a]Center of Excellence in Design, Robotics and Automation (CEDRA),Department of Mechanical Engineering,
Sharif University of Technology, Tehran, Iran*
sakhaee@mech.sharif.ir

Abstract. A finite element technique is used to mimic radial deformation of single-walled carbon nanotubes under hydrostatic pressure. The elastic deformation of nanotubes is modeled via elastic beams. Properties of the beam element are evaluated by considering characteristics of the covalent bonds between the carbon atoms in a hexagonal lattice. Applying the beam model in a three dimensional space, the elastic properties of the nanotube in the transverse direction are evaluated. The effects of diameter and wall thickness on the radial and circumferential elastic moduli of zigzag and armchair nanotubes are considered. Results are in good agreement with molecular structural mechanics data in the literature.

Keywords: A. Single-walled carbon nanotubes (SWCNT); Finite element analysis (FEA); B. Elastic properties.
PACS: 62.25. +g

INTRODUCTION

Carbon nanotubes (CNTs) were discovered in 1991 by Ijima [1]. Extraordinary characteristics of the nanotube are so promising that many researchers have attempted to develop and enhance the mechanical behavior of these materials. The essential requirement of developing a reliable mechanical behavior of CNTs occurred when the motivation to use these materials appeared.

Mechanical characteristics of CNTs have been investigated by experimental observation and theoretical prediction [2-5]. Although mechanical properties in different directions have been reported, a growing effort in modeling of the carbon nanotube behavior can be found in the literature.

The elastic characteristics of SWCNT were tackled via an elaborate analytical approach [6]. Naghdabadi et al. [7] proposed an analytical approach of bending modulus of multi-layered graphene sheets which may be implemented as an idealized large CNT. An analytical molecular structural approach to simulate the radial deformation has also been reported by Xiao and co-workers [8].

Since the analytical approaches are elaborate, an inclination toward numerical techniques appeared. The material properties of nanotube were studied via a square representative volume element [9]. Behavior of CNT under bending was modeled using shell element [10].

In another study, the finite element technique was implemented to determine stiffness and strength of single-walled carbon nanotubes (SWCNT) while considering inter atomic forces [11]. A new four-node element was proposed by Nasdala and Ernst for nano materials [12]. Meo and Rossi developed and implemented a new non-linear spring element to calculate Young's modulus of elasticity [13].

Li and Chou [14] proposed a beam model which has a capability of representing the covalent bond characteristics. By introducing the beam model the aim of simple modeling of CNTs behavior using the FE technique was achieved.

Tserpes and Papanikos [15] implemented the beam model to mimic the SWCNT behavior under tension and torsion. Their results exhibit good agreement within a maximum 3 percent difference with respect to stiffness approach [14].

CP929, *Nanotechnology and Its Applications, First Sharjah International Conference*
edited by Y. I. Salamin, N. M. Hamdan, H. Al-Awadhi, N. M. Jisrawi, and N. Tabet
© 2007 American Institute of Physics 978-0-7354-0439-7/07/$23.00

They also investigated the effect of SW defects on fracture of nanotubes [16]. Cho[17] reported the bending and shear modulus of SWCNT while considering the Poisson's effect. Kalamkarov et al [18] also used the FE method to model the single-walled and multi-walled carbon nanotubes under tension and torsion.

In this study, by applying the beam model in a three dimensional coordinate system, the radial deformation, of zigzag and armchair single-walled carbon nanotubes under hydrostatic pressure, is studied.

To obtain beam element properties, the characteristics of the covalent bonds between the carbon atoms in a hexagonal lattice are considered. By considering small deformation in determining the elastic characteristics, modeling with linear element would suffice.

SWCNT ATOMIC STRUCTURE

Carbon nanotubes are assumed to be made by rolling a graphene sheet about the $\vec{T}$ vector. Perpendicular to the $\vec{T}$ vector is defined as the chiral vector and is denoted by $\vec{C}_h$. Using base vectors $\vec{a}_1$ and $\vec{a}_2$, the chiral vector is written as

$$\vec{C}_h = n * \vec{a}_1 + m * \vec{a}_2 \qquad (1)$$

where n and m are transformation indices and are an indication of the structure of the zigzag or armchair numbers in the circumferential path. Due to the varying structures of carbon nanotubes, they exhibit different mechanical properties. In this investigation, the radial deformation of SWCNT with zigzag and armchair are analyzed.

The chiral vector can be best introduced by lattice transformation indices (n, m), where base vectors are shown in Fig. 1.

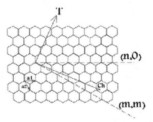

FIGURE 1. Hexagonal lattice of graphene sheet including base vectors of SWCNT.

By employing lattice transformation indices, the chiral vector can be defined by (n, 0) for zigzag and (m, m) for armchair.

FINITE ELEMENT MODELING OF SWCNT

In order to mimic the radial deformation of SWCNTs under hydrostatic pressure, a model, consisting of nanotube beam elements, is considered. Beam element elastic properties in the FE model of the SWCNTs are evaluated by considering molecular and structural potential energy, based on the proposed Odegard technique [19].The model is clarified as follow.

By omitting the electrostatic interaction for covalent systems under small deformation, total steric potential energy can be expressed as [20]

$$U_{total} = \sum U_r + \sum U_\theta + \sum U_\tau , \qquad (2)$$

where U_r is the bond stretch energy and can be written as

$$U_r = \frac{1}{2} k_r (r - r_o)^2 = \frac{1}{2} k_r (\Delta r)^2 . \tag{3}$$

Bond angle variation results in bending energy and can be calculated as

$$U_\theta = \frac{1}{2} k_\theta (\theta - \theta_o)^2 = \frac{1}{2} k_\theta (\Delta\theta)^2 . \tag{4}$$

In Eq. 2, U_τ denotes torsional energy and can be evaluated as

$$U_\phi = \frac{1}{2} k_\phi (\Delta\phi)^2 . \tag{5}$$

Here, k_r, k_θ and k_ϕ represent bond stretching ,bond angle bending and torsional stiffness. Δr, $\Delta\theta$ and $\Delta\phi$ are the corresponding increments. It is clear that the potential energies of different skims, based on molecular mechanics, are independent.

According to structural mechanics, the strain energy of a uniform beam under axial load, N, can be found as

$$U_A = \frac{1}{2} \int_0^L \frac{N^2}{EA} dL = \frac{1}{2} \frac{N^2 L}{EA} = \frac{1}{2} \frac{EA}{L} (\Delta L)^2, \tag{6}$$

where A is the cross section of the beam, L stands for beam length, ΔL denotes its variation and E represents Young modulus of the considered beam .Due to pure bending load, M, the strain energy in the beam, can be determined as

$$U_M = \frac{1}{2} \int_0^L \frac{M^2}{EI} dL = \frac{2EI}{L} \alpha^2 = \frac{1}{2} \frac{EI}{L} (2\alpha)^2, \tag{7}$$

where I is the moment of inertia and α stands for the rotational angle of the beam ends. Pure torsional loading results in strain energy that can be written as

$$U_T = \frac{1}{2} \int_0^L \frac{T^2}{GJ} dL = \frac{1}{2} \frac{T^2 L}{GJ} = \frac{1}{2} \frac{GJ}{L} (\Delta\beta)^2, \tag{8}$$

where $\Delta\beta$ denotes the relative rotation angle of two ends, G is the shear modulus and J represents the polar moment of inertia .No interaction between different types of stored energy in structural mechanics and molecular mechanics can be found .

Due to the independency of energy terms in each system; i.e. molecular and structural mechanics, the equivalency of corresponding types in two systems can be imposed. The energy equivalency between different types of potential energy in molecular and structural mechanics, as shown, indicates that

$$\frac{EA}{L} = k_r, \frac{EI}{L} = k_\theta, \frac{GJ}{L} = k_\phi, \tag{9}$$

EA, EI and GJ imply beam element characteristics. The properties of a beam element with a circular cross section can be calculated, as follow [9]

$$d = 4\sqrt{\frac{k_\theta}{k_r}}$$

$$E = \frac{k_r^2 L}{4\pi k_\theta} \quad , \tag{10}$$

$$G = \frac{k_r^2 k_\phi L}{8\pi k_\theta^2}$$

where d, E and G represent cross section diameter, modulus of elasticity and shear modulus of beam element. L is the element beam length, equal to the covalent bond length of carbon atoms in a hexagonal lattice.

RESULTS

To mimic hydrostatic pressure, radial point forces are applied to the nodes of bridged nanotube. The nodes of the frame-structure model are assumed to be coincident with the carbon atoms. In the proposed FE model, the equivalent hydrostatic pressure can be evaluated as

$$P = \frac{nf}{2\pi Rl} , \tag{11}$$

where f, n and l are the magnitude of radial point forces, number of the nodes and length of the space-frame structure. Radial and hoop stresses in the model can be calculated as

$$\sigma_r = P$$

$$\sigma_\theta = \frac{PR}{t} , \tag{12}$$

in which P and t are the equivalent hydrostatic pressure and tube wall thickness. Stimulated strains of the hydrostatic pressure can be derived as

$$\varepsilon_r = \varepsilon_\theta = \frac{u_r}{R} , \tag{13}$$

where R and u_r are the nanotube radius and radial displacement. Radial displacements are considered far from the ends where displacement assumed to be constant and implemented in the measurement of the elastic properties. Neglecting the Poisson's effect, transverse elastic moduli can be defined as

$$E_R = \frac{PR}{u_r} \text{ and } E_\theta = \frac{PR^2}{tu_r} , \tag{14}$$

with the molecular dynamics constants given in Table. 1, the radial and circumferential moduli of elasticity are derived.

To model the space-frame structure of SWCNT, the ANSYS commercial package is used. The elastic properties of the beam element are applied to the beam (3D elastic BEAM4). The radial deformation of zigzag SWCNT under hydrostatic pressure is demonstrated in Fig. 2.

FIGURE 2. Radial deformation of zigzag structure under hydrostatic pressure.

The radial strain of bridged zigzag and armchair nanotube under hydrostatic pressure from the proposed model and molecular structural method are shown in Fig.3. Due to absence of any non-linear interaction in this model, linear dependency of the strain to the hydrostatic pressure [21] is clear.

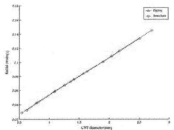

FIGURE 3. Stimulated Radial strain under hydrostatic pressure (P = 37.8 Mpa).

The elastic properties of the carbon nanotubes in transverse direction from the present model and the given data in the literature [21] are presented in Figs. 4 and 5. Comparing the results indicates good agreement within a maximum 5 percent difference.

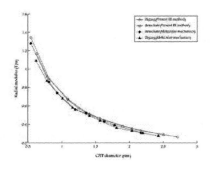

FIGURE 4. Radial direction modulus of elasticity of SWCNT.

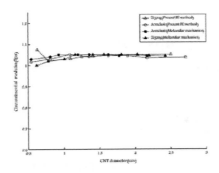

FIGURE 5. Circumferential direction modulus of elasticity of SWCNT.

It is seen that chirality plays an insignificant role in the transverse elastic characteristics. Numerical results depict that increasing diameter results in a lower radial modulus while not having a notable effect on the circumferential modulus.

In the present FE method, Eq. (14) depicts increasing tube wall thickness has no effect on the radial modulus of elasticity while the circumferential modulus is inversely proportional to it. This is in good agreement with previous studies of the longitudinal modulus [19, 23].

Promising results and their agreement with a molecular mechanics approach ensure the precision of the method while simplicity and time saving in the analysis are achieved.

CONCLUSION

In this paper, radial deformation of single-walled carbon nanotube under hydrostatic pressure is modeled using a 3-dimensional FE beam. The Effects of diameter and chirality on the radial deformation are considered. Obtained finite element data is used to calculate elastic properties of CNTs.

Numerical results exhibit circumferential modulus of elasticity to be insensitive with respect to diameter while radial modulus of elasticity decreases non-linearly with increasing diameter. Since chirality has insignificant role in transverse elastic characteristics, one formula for radial modulus is developed. This model depict increasing tube wall thickness has no effect on radial modulus of elasticity while circumferential modulus is inversely proportional to it. Based on negligible effect of chirality, an empirical equation for the prediction of radial modulus of elasticity is developed.

Due to precise assessment of mechanical behavior in tension, torsion, bending and radial expansion under hydrostatic pressure, the proposed FE method is highly appreciated. Simplicity while being computationally efficient is the main achievement of the implemented technique.

REFERENCES

1. Iijima, S., Nature **354** (1991) 56–8.
2. Treacy, M. M. J., Ebbesen, T. W., Gibsin, J. M., Nature **381** (1996) 678-80.
3. Krishhan, A., Dujardin, E., Ebbesen, T. W., Yianilos, P. N., Treacy, M. M. J., Phys. Rev B **58** (1998) 14013-9.
4. Ruoff, R. S., Qian, D., Liu, W. K., CR Phys. **4** (2003) 993-1008.
5. Lau, K. T., Chipara, M., Ling, H. Y., Hui, D., Compos. Part B **35** (2003) 95-101.
6. Natsuki, T., Tantrakarn, K., Endo, M., Carbon **42** (2004) 39–45.
7. Behfar, K., Seifi, P., Naghdabadi, R., Ghanbari J., J. Thin Solid Films **496** (2006) 475-480.
8. Xiao, J. R., Gama, B. A., Gillespie, Jr., Int. J. Solids Struct. **42** (2005) 3075–92.
9. Chen, X. L., Liu, Y. J., Computational Materials Science **29** (2004) 1–11.
10. Pantano, A., Parks, D. M., Boyce, M. C., Journal of the Mechanics and Physics of Solids **52** (2004) 789 – 821.

11. Sun X., Zhao W., Mater Sci Eng A 390(2005) 366–71.
12. Nasdala, L., Ernst G. Computational Mater, Sci. **33** (2005) 443–458.
13. Meo, M., Rossi, M., Compos Scie and Tech. **66** (2006) 1597–1605.
14. C. Li. And T. W. Chou, Int J. Solids Struct. **40** (2003) 2487–99.
15. Tserpes, K. I., Papanikos, P., Compos: Part B. **36** (2005) 468–477.
16. Tserpes, K. I., Papanikos, P., Compos. Sruc. (in press).
17. Cho, W. S. To, Finite Elements in Analysis and Design. **42** (2006) 404-413.
18. Kalamakov, A. L., Georgiades, A. V., Rokkam, S. K., Veedu, V. P., Ghasemi-Nejhad, M. N., Int J Solids Struct (in press).
19. Odegard, G. M., Gates, T. S., Nicholson, L. M., Wise, K. E., Compos. Sci. Technol. **62** (2002) 1869–80.
20. Gelin B. R., Molecular Modeling of Polymer Structures and Properties, (Hanser/Gardner Publishers, Cincinnati, OH, 1994).
21. Li C Y., Chou T W., Phys. Rev. B **69** (2004) 073401.
22. Mendenhall W., Reinmuth J E., Beaver R., Statistics for Management and Economics, 6th edition (PWS-Kent Publishing Company, Boston, Massachusetts, 1989).
23. Gao X. L., Li K., Int J. Solid Struct. **40** (2003) 7329-39.

Investigation of Boron and Nitrogen-doped CNTs by Dispersive Micro-Raman Back-Scattering Spectroscopy

R. Malekfar[1], S. Hosseini[1], B. Fakhraei[2], M. Khanlary[2], H. Asadi[1] and S. Babanejad[3]

[1]Physics Department, Faculty of Basic Sciences, Tarbiat Modarres University, P.O. Box 14115-175, Tehran, Iran
Malekfar@Modares.ac.ir
[2]Physics Department, Imam Khomeini International University, Qazvin, Iran
[3]Physics Department, Payam-e Noor University of Sari, Sari, Iran

Abstract. Raman scattering studies reveal the remarkable structure and the unusual electronic and phonon properties of carbon nanotubes, CNTs. In this study, we directly produced boron-, B, and nitrogen-, N, doped CNTs by using the DC-arc discharge method which normally can be employed for producing CNTs. We performed a series of experiments with and without Fe element as the catalyst and in the presence of C_2H_2 gas for producing boron doped CNTs. At the second stage and in the presence of an Fe catalyst, C_2H_2 gas and nitrogen, N_2, gas were used for producing nitrogen doped CNTs. In general, our investigation revealed that some major changes caused by B and N dopants can be observed in the related recorded Raman spectra.

Keywords: Raman spectroscopy, B and N doped CNTs, DC-Arc discharge method.
PACS: 78.67.Ch, 78.30.-j.

INTRODUCTION

Raman spectroscopy is a powerful tool for characterizing the unique optical properties of carbon nanotubes. Raman scattering reveals structural and unusual electronic, phonon properties and metallic and semiconducting phases of carbon nanotubes. This investigation has proven to be a very useful probe for carbon-based materials and has been used extensively to study the bonding, semi-conducting and superconducting phases of CNTs and also changes caused by different doping levels. In this study, we have investigated the B and N doped CNTs produced by arc discharge method, using the dispersive back scattering micro Raman spectroscopy.

CARACTERISTICS OF RAMAN SPECTRA OF CNTS

Basically, the most common characteristics of Raman peaks of CNTs are as follow:
1. A peak, or sometimes peaks or a band, at low frequencies which is called Radial Breathing Mode, RBM, which strongly depends on the nanotubes diameters and the singularities of the tubes.
2. A peak or a band with relatively higher frequency caused by disorders which is known as the D-line.
3. A peak, usually divided into two peaks, or bands at higher frequencies related to the atomic vibrations along the circumferential direction and along their axis direction, respectively, which are called G^+ and G^- and are seen clearly in semiconducting tubes.
4. A peak or a band observed as the first overtone of the D-band which is mostly defined by G' (Fig. 1).

In our study, we directly produced boron, B, and nitrogen, N, doped CNTs by using DC-arc discharge method which normally can be employed for producing CNTs. First, we performed a series of experiments with and without Fe

CP929, Nanotechnology and Its Applications, First Sharjah International Conference
edited by Y. I. Salamin, N. M. Hamdan, H. Al-Awadhi, N. M. Jisrawi, and N. Tabet
© 2007 American Institute of Physics 978-0-7354-0439-7/07/$23.00

element as the catalyst and in the presence of C_2H_2 gas for producing boron doped CNTs. At the second stage, in the presence of Fe catalyst, C_2H_2 gas and nitrogen, N_2, gas were used for producing nitrogen doped CNTs.

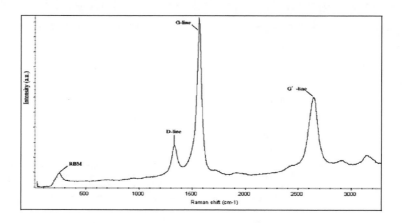

FIGURE 1. Characteristic Raman peaks of CNTs in the spectral region of 40-4000 cm^{-1}.

Doping Effects

Boron and nitrogen are the nearest elements to carbon in the periodic table, doping of graphite with such elements will modify the electronic, mechanical and oxidative properties of these materials and offer the opportunity not only to understand dopands induced perturbation in physical properties of one dimensional materials but such doping also provides an opportunity to exploit their unique properties in the next generation of doping technologies with nitrogen and boron which are respectively expected to behave as a donor (n-type) or acceptor (p-type) dopands and enhance CNTs conductivity.

EXPERIMENTS

The doped carbon nanotube samples used in this study were prepared by a DC arc discharge method. Primarily, anode electrode was a graphite rod with 4.5 mm diameter and 3cm length which was drilled for constructing a hole with 2.25 mm diameter and depth with a ratio of 2/3 of electrode length. Cathode electrode also was a graphite rod with 2.5 mm diameter and 3 cm in length.

In the first section of our experiments and for producing boron doped carbon nanotubes we used C_2H_2 atmosphere as the hydrocarbon source and, without using catalyst particles, we filled the hole of the anode with a suspension of ethanol solution and boron powder at the C/B ratio of 2/1 by weight. The dark grey filamentous material deposited on the cathode electrode was collected and dissolved in boric acid for 18 hour and washed with de-ionized distilled water. After filtering the samples, they were heated in an oven at temperature from 400°C to 450°C for 4 hours.

In the second section, we produced boron doped carbon nanotubes in C_2H_2 atmosphere by using Fe element as catalyst powder at Fe/B ratio of 1/3 by weight (1% Fe, 3% B and 96% graphite). A DC current of 30 ampere in a gas flow of 20 sccm of C_2H_2 was employed. The samples deposited on the cathode electrode were collected and dissolved in nitric acid for 18 hour and washed with de-ionized distilled water. For drying the samples, they were heated in an oven at temperatures of about 400°C to 450°C for 2 hours and after being dissolved in boric acid for 18 hours and washed with de-ionized water, they were heated again in the oven from 400°C to 450°C for another 2 hours.

In the next stage and for producing nitrogen doped carbon nanotubes iron as catalyst at C_2H_2-N_2 atmosphere were used. The related molecular weight of iron and graphite was 3% iron and 97% graphite in ethanol alcohol suspension in a DC current varied from 30 to 200 A. The samples were purified for Raman spectroscopy and the best synthesized samples were produced at 200 A for nitrogen doped CNTs under DC current of 30 A and a gas flow of 350 sccm of N_2 gas.

RESULTS AND DISCUSSIONS

In Figs. 2 and 3, spectra of pure and doped CNTs in the absence and presence of iron and boron catalysts at C_2H_2 atmosphere have been presented. In Fig. 4 typical SEM images of nitrogen and boron doped produced CNTs in our laboratory are shown. As clearly can be seen from the Raman spectra the N doped Raman modes shift to lower wave-numbers, whereas the CNTs doped by acceptor reagents such as B were found to shift to higher wave-numbers.

In the former case, the related N dopants act as an electron donor. The extra electron tends to weaken the C-C bonds in the CNTs because electrons have been known to soften the C-C bond in all sp^2-bonded carbon material. The results of weakened bond are a downshift in tangential band to lower frequencies.

However, in the case of boron doping, the Raman spectra of the intercalated carbon nanotubes show opposite behavior and towards the higher wave-numbers and the dopants are expected to act as an electron acceptor.

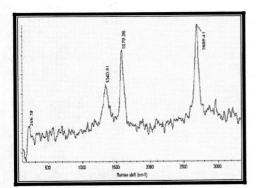
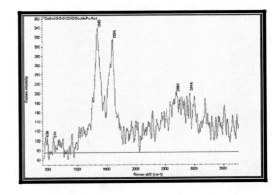

FIGURE 2. Characteristic Raman peaks of pure carbon nanotubes with iron catalyst, left spectrum, and boron doped carbon nanotubes without iron catalyst, right spectrum.

91

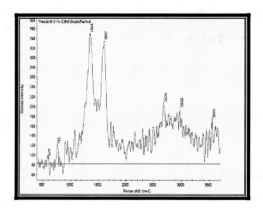 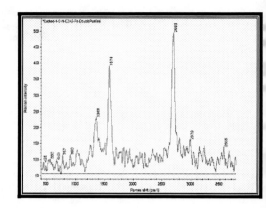

FIGURE 3. The Raman spectra of boron doped carbon nanotubes with iron catalyst, left spectrum, and nitrogen doped carbon nanotubes with iron catalyst, right spectrum.

FIGURE 4. Typical SEM images of nitrogen doped CNTs, left image, and boron doped CNTs, right image.

REFERENCES

1. M. Terrones, A. Jorio, M. Endo, A. M. Rao, Y. A. Kim, T. Hayashi, H. Terrones, J.-C. Charlier, G. Dresselhaus, and M. S. Dresselhaus, Els. Materialstoday, October (2004).
2. A. Jorio, M. A. Pimenta, A. G. Souza Filho, R. Saito, G. Dresselhaus & M. S. Dresselhaus, Characterizing carbon nanotube samples with resonance Raman scattering, New Journal of Physics **5**, 139.1-139.17 (2003).
3. S. H. Lim, H. I. Elim, X. Y. Gao, A. T. S. Wee, W. Ji, J. Y. Lee & J. Lin, Electronic and optical properties of nitrogen-doped multiwalled carbon nanotubes, Physical Review B **73**, 045402 (2006).
4. M. Meyyappan (Editor), Carbon Nanotubes (Science & Applications), CRC Press, (2004).

SYNTHESIS AND CHARACTERIZATION

Morphology Tuning of Strontium Tungstate Nanoparticles

S. Joseph[*], T. George[*], K. C. George[†], A. T. Sunny[¶] and S. Mathew[¶]

[*]Nirmala College, Muvattupuzha, 686 661, Kerala, India
[†]S.B. College, Changanacherry, 686 101, Kerala, India
[¶]School of Chemical Sciences, Mahatma Gandhi University, Kottayam, 686 560, Kerala, India
smathew_mgu@yahoo.com

Abstract. Strontium tungstate nanocrystals in two different morphologies are successfully synthesized by controlled precipitation in aqueous and in poly vinyl alcohol (PVA) medium. Structural characterizations are carried out by XRD and SEM. The average particle size calculated for the $SrWO_4$ prepared in the two different solvents ranges 20-24 nm. The SEM pictures show that the surface morphologies of the $SrWO_4$ nanoparticles in aqueous medium resemble mushroom and the $SrWO_4$ nanoparticles in PVA medium resemble cauliflower. Investigations on the room temperature luminescent properties of the strontium tungstate nanoparticles prepared in aqueous and PVA medium shows strong emissions around 425 nm.

Keywords: nanostructures; soft chemistry; luminescence; strontium tungstate.
PACS: 61.46.Df; 68.37Hk; 73.63.Bd

INTRODUCTION

Nanostructured tungstate materials have aroused much interest because of their luminescence behavior, structural properties and potential applications in the fabrication of photoelectric materials. It is instructive to study the synthesis and physical properties of strontium tungstate ($SrWO_4$), a candidate for the design of solid state laser, which is compositionally and structurally a representative of the ABO_4 class of metal oxides where A and B are two different metallic elements with +2 and +6 oxidation states. Pure and rare earth-doped strontium tungstates have received a great deal of interest in the past decades in the fields such as electron paramagnetic resonance, dielectric properties, microstructure, Raman spectra characteristics and Raman self conversion media [1-4]. The tungstate crystals with scheelite structure have very similar spectral characteristics (AWO_4, A = Pb, Ca, Ba, Sr). Scheelite type $SrWO_4$ has been widely used in the optoelectronic industry and the solid-state laser system due to its luminescence behaviour and stimulating Raman scattering property [5]. Different methods for the synthesis and morphology control of tungstate nano crystals have been reported [6-9]. At high temperature there is a tendency for the WO_3 group to evaporate, resulting inhomogeneous composition of tungstates. Phosphors prepared by wet chemical method have higher uniformity in particle size distribution with good crystallanity and exhibit higher photoluminescent intensity than those of the solid-state reaction prepared [10]. In this investigation we have adopted a simple and highly reproducible precipitation method for the synthesis of nanosized $SrWO_4$ in two morphologies, by changing the solvents (distilled water and PVA).

EXPERIMENTAL

Strontium nitrate ($Sr(NO_3)_2$) and sodium tungstate ($Na_2WO_4 . 2H_2O$) with analytical purity (99.9%), purchased from Aldrich Sigma were dissolved in distilled water with a concentration of 0.05M each. The solutions were mixed and stirred vigorously to ensure homogeneous dispersal of reagents in the solution. White precipitate formed were collected, washed repeatedly with distilled water and ethanol and dried at 100 °C. Poly vinyl alcohol (PVA) solution at a concentration of 4 g L^{-1} was prepared in two beakers and strontium nitrate and sodium tungstate (0.05M each) were dissolved in the PVA solutions and the above procedure was repeated. The X-ray diffraction

CP929, *Nanotechnology and Its Applications, First Sharjah International Conference*
edited by Y. I. Salamin, N. M. Hamdan, H. Al-Awadhi, N. M. Jisrawi, and N. Tabet
© 2007 American Institute of Physics 978-0-7354-0439-7/07/$23.00

patterns of the powders were recorded using 'Bruker D8 Advance' X-ray diffractometer using Cu K_α radiation (λ = 1.5406 Å). The morphology of the strontium tungstate nanostructures was examined with scanning electron microscope (JEOL-JSM-5600 LV). UV-vis spectra were taken using a UV–visible spectrophotometer (Schimadzu UV-2400 PC) with a reflection mode attachment. The luminescence studies were performed at room temperature using 'Spex Flouromax 3-Flourimeter'.

RESULTS AND DISCUSSION

Figure 1 shows the X-ray diffraction patterns of $SrWO_4$ nanoparticles synthesized in (a) aqueous medium and (b) PVA medium. The samples are well crystallized and the patterns agree well with the JCPDS card No: 08-0490 and the crystals are scheelite tetragonal with space group $I4_1/a$. The average crystallite size calculated from the full width at half maximum (FWHM) of the major peaks using the Scherrer formula is 24 nm for the sample (a) and 20 nm for sample (b).

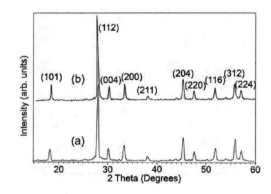

FIGURE 1. X-ray powder diffraction of patterns of strontium tungstate nanoparticles synthesized in (a) distilled water (b) PVA.

Figure 2 illustrates the morphology of strontium tungstate nanostructures examined with scanning electron microscope (SEM). The SEM pictures show that the $SrWO_4$ nanoparticles synthesized in distilled water medium resemble mushroom (Fig. 2a, 2b) and the $SrWO_4$ nanoparticles in PVA medium resemble cauliflower (Fig. 3a, 3b). It is well known that definite separation of nucleation and growth stages are the primary requirements for uniform particle formation. Generally the shape of a crystal is determined by the relative specific surface energies associated with the facets of the crystal. It is a well-established fact that the media plays a crucial role in the growth process of nanoparticles [8]. In PVA medium, the absorbed OH^- ion may have controlled the growth rate in various crystal faces. The exposed faces of $SrWO_4$ crystals show different polarity patterns and averaged interface energies thus adsorb the medium (PVA) with different efficiencies. The PVA modified crystal would grow preferentially in the direction where crystallization hindrance is the weakest. Thermodynamically all of the nanocrystals will grow towards the shape having the lowest energy at equilibrium. Small primary particles are formed right after the mixing of the reactants at the molecular level. These particles further aggregate in some specific orientations.

FIGURE 2. SEM images (a, b) of strontium tungstate nanoparticles synthesized in distilled water.

The UV-visible spectrum (Fig. 4) $SrWO_4$ nanoparticles synthesized in aqueous medium shows an absorbance maximum at 235 nm. This absorption is attributed to a charge transfer transition in which an oxygen 2p electron goes into one of the empty tungsten 5d orbital. Due to UV irradiation of strontium tungstate, excitation from O_{2p} to W_{t2g} in the WO_4^{2-} group occurs. In the excited state of the WO_4^{2-} groups, the hole (on the oxygen) and the electron (on the tungsten) remain together as an exciton because of their strong interactions.

Investigations on the room temperature luminescent properties (Fig. 5) of the strontium tungstate nanostructures synthesized in aqueous medium and in PVA medium (excitation wavelength, $\lambda_{ex} = 248$ nm) display weak emissions around 280 nm and strong emissions around 425 nm, due to the regular lattice. The numerous crystal defect centers (Frenkel defects) relative to oxygen also contributes much to the luminescence spectrum. The metal tungstates exhibit the blue luminescent spectra, which are based on the radiative transition within the tetrahedral (WO_4^{2-}) group [11-12]. It is found that the PL emission of the sample synthesized in PVA medium (Fig. 3b) is more intense than the sample synthesized in distilled water medium (Fig. 3a). The increase in intensity may be due to the decrease in particle size and the increase in surface defects.

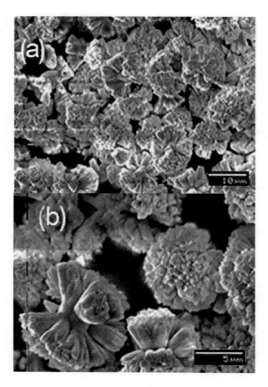

FIGURE 3. SEM images (a, b) of strontium tungstate nanoparticles synthesized in the medium PVA.

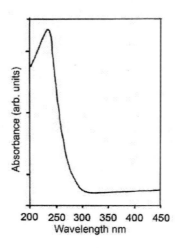

FIGURE 4. UV-vis spectrum of $SrWO_4$ nanoparticles synthesized in aqueous medium.

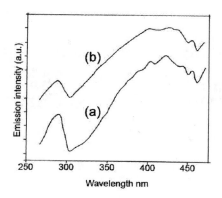

FIGURE 5. Photoluminescence (PL) spectrum (λ_{ex} = 248 nm) of strontium tungstate nanoparticles synthesized in (a) distilled water (b) PVA.

ACKNOWLEDGMENTS

The authors are grateful to SAIF, Cochin University, Kochi, RRL- Trivandrum and IISc (SSCU), Bangalore for characterization.

REFERENCES

1. J. P. Sattler and J. Nemarich, Phys. Rev. B **1**, 4249- 4256 (1970).
2. X. Zhao, T. L. Y. Cheung, X. Zhang and D. H. L. Ng, J. Am. Ceram. Soc. **89**, 2960-2963 (2006).
3. J. Y. Huang and Q. X. Jia, Thin Solid films **444**, 95-98 (2003).
4. S. P. S. Porto and J. F. Scott, Phys. Rev. **157**, 716- 719 (1967).
5. Z. D. Lou and M. Cocivera, Mater. Res. Bull. **37**, 1573-1582 (2002).
6. D. Chen, G. Shen, K. Tang, Z. Liang and H. Zheng, *J. Phys. Chem. B* **108**, 11280-11284 (2004).
7. A. Dias and V. S. T. Ciminelli, Chem. Mater. **15**, 1344-1352 (2003).
8. J. Geng, Y. Lv, D. Lu and J-J Zhu, *Nanotechnology* **17**, 2614-2620 (2006).
9. T. George, S. Joseph and S. Mathew, Pramana- J. Physics **65**, 793-799 (2005).
10. W. S. Cho, M. Yashima, M. Kakihana, A. Kudo, T. Sakata, and M. Yoshimura, *Appl. Phys. Lett.* **66**, 1027-1029 (1995).
11. M. Nikl, *Phys. Stat. Sol.* **178**, 595-620 (2000).
12. G. Zhang, R. Jia and Q. Wu, Mater. Sci. Eng. B **128**, 254-259 (2006).

Formation of a Conductive Nanocomposite on Plastic and Glass Substrates Through Wet Chemical Deposition of Well Dispersed Carbon Nanotubes

N. Al-Dahoudi

Physics Department-Al Azhar university-Gaza, P. O. Box 1277, Palestine
naji@alazhar-gaza.edu

Abstract. Carbon multi-wall nanotubes (MWNTs) powder was successfully dispersed using microfluidizer with different loading ratio in a water-based inorganic organic monomer composing a stable coating sol. The coatings made of the sol showed that the conductivity behave according to a power percolation law model with critical volume fraction of 0.0825. The highest obtained electrical conductivity of a single 85 nm thick layer of the system was 10^3 $\Omega^{-1}.m^{-1}$. At the same time the coatings are transparent showing a transmission quite similar to that of the substrate.

Keywords: Transparent conductive coating, carbon Nanotubes, sol gel, percolation.
PACS: 61.64.FG, 73.63.FG

INTRODUCTION

Systems composed of a mixture of conducting and insulating species have gained an increasing interest due to their applications in thin film technology, e.g. for electromagnetic shielding and antistatic purposes [1-4]. The electrical properties of such systems mainly depend on the dispersion condition of filler particles, particle size, the aggregate structure and the fraction of the conducting species in the system [5]. These systems have poor conductivity below certain volume fraction of the conducting filler particles called the critical threshold, where there are no contacts between adjacent filler particles. Above the critical threshold, agglomerates are grown to reach a size which makes it possible for them to touch each other resulting in a formation of a conductive network within the insulating phase. To understand the network formation, there are many so-called percolation models (6).

Carbon nanotubes are very interesting materials which drive remarkable scientific research in a number of areas including chemistry, electronic transport, mechanical and field emission properties. The use of carbon nanotubes to impart electrical conductivity, while maintaining high optical transparency, is nowadays possible [7, 8]. The unique quality of this form of carbon is simple in that the individual particles possess the attributes of high electrical conductivity and high aspect ratio combined with a unique capability of forming ropes of individual particles. The combination of all these properties allows the formation of conductive networks through the material with tunable electrical resistivity and good transparency. Conductive colloid, on which carbon multiwall nanotubes are dispersed, can be coated on different substrates by wet deposition techniques, which is useful when large surface area coatings with low cost are needed.

EXPERIMENTAL

Aqueous Suspensions containing carbon MWNTs as a conducting filler and a water based epoxy functionalized silane were prepared to provide a range of carbon MWNTs loading in a water lack

CP929, *Nanotechnology and Its Applications, First Sharjah International Conference*
edited by Y. I. Salamin, N. M. Hamdan, H. Al-Awadhi, N. M. Jisrawi, and N. Tabet
© 2007 American Institute of Physics 978-0-7354-0439-7/07/$23.00

matrix. First a very thin functionalized carbon MWNTs powder was dispersed using a dispersion agent in water based epoxy functionalized alkoxysilane by mixing them under stirring in an ultrasonic bath. A microfluidizer offers a high shear fluid processing was used to disperse large agglomerates into uniform stable dispersions, which provide a range of carbon MWNTs loading, from 0.05 to 0.45 weight fraction (in the dried coating). The coating sols were deposited on glass or polycarbonate (PC) substrates using the spin coating technique. The wet layers were dried for 10 min at 40 °C and then heated in an oven for 30 min at temperatures range between 80 and 400 °C.

The electrical resistivity (ρ) was measured using a contactless system from Lehighton Electronics, Inc. A white light interferometer (WLI) from Zygo New View 500 was used for the analysis of the surface morphology. A CARY 5E UV-VIS-NIR spectrophotometer was used to measure the transmission of the coatings.

RESULTS AND DISCUSSION

Electrical Properties

The electrical conductivity behavior of layers deposited on PC substrate using spin coating method dried for 10 min and then heated at 130 °C for 30 min is shown in Fig.(1) as a function of the loading volume ratio of carbon MWNTs in the insulating matrix. In the region of filler low concentration curve, the system exhibits a poor conductivity and at certain concentration between 0.06 and 0.1 a sharp increase of the conductivity is observed. The highest obtained electrical conductivity of the system is on the order of $10^3 \ \Omega^{-1} m^{-1}$.

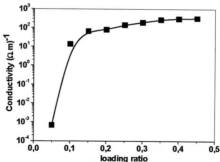

FIGURE 1. The electrical conductivity of spin coated nanocomposite layer on PC substrate as a function of the loading ratio of the carbon MWNTs.

To understand the behavior shown previously, a classical model is proposed to explain the electrical conductivity of mixtures made of conductive and insulating material [9], in which

$$\sigma. = \sigma_0 (V - V_c)^s, \tag{1}$$

where σ is the direct-current conductivity, σ_0 is the proportionality constant which represents the intrinsic electrical conductivity of the conductive filler, V is the conductive filler volume fraction, V_c is the percolation threshold at which the conductivity of the composite starts to increase sharply, and s is the critical conductivity exponent.

By applying the percolation power law equation to our experimental data of Fig. 1, it shows that the data could be well fitted as shown in Fig. 2. The volume fraction $V_c = 0.0825$ was chosen as the percolation threshold which results in the best linear fit with R value of 0.993. So, the conductivity behavior of the carbon MWNTs composite coatings agrees well with the percolation power law showing a lower percolation threshold.

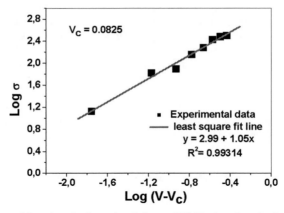

FIGURE 2. A log-log plot of the electrical conductivity and V-V$_c$ showing the least square-fit line for V$_c$ =0.0825

Such behavior is attributed to the contact between the adjacent filler tubes. In the region of V$_c$ < 0.0825 there are no contacts between the nanotubes and the conductive filler distributed within the insulating species. By raising the nanotube concentration above the percolation threshold, the carbon nanotubes begin to agglomerate and touch each other forming a conductive network leading to a drastic increase of the electrical conductivity of the coating. By further increment of the conductive nanotubes, the electrical conductivity shows a slow increase, which is attributed to better improvement of the quality of the conductive network. Figure 3 shows WLI images for the surface of carbon MWNTs composite coatings with different loading ratio. Figure 3-a shows a WLI surface image for coating containing only 0.05 volume fraction of carbon MWNTs loading. It appears as an isolated cluster, which is different from the 0.1 volume fraction shown in Fig. 3-b, where an interconnected network of the nanotubes encapsulating within the insulating matrix appears on the surface. This interconnection explains the drastic increase of the electrical conductivity. Figure 3-c shows the same image for a 0.3 volume fraction, where more homogeneous network of nanotubes were formed enhancing the electrical conductivity of the coatings.

Figure 4 shows the electrical resistivity of the carbon MWNTs composite coatings on a glass substrate as a function of the heating temperature. The electrical resistivity of the layer decreases by increasing the heating temperature, where better contact between the conducting filler occurred lowering the pores in the layer. A minimum electrical resistivity of 1.5×10^{-3} Ω.m is obtained at heating temperature of 300 °C. An increase of the resistivity occurs at heating temperature above 400 °C, suggesting a change on the carbon MWNTs properties at this temperature.

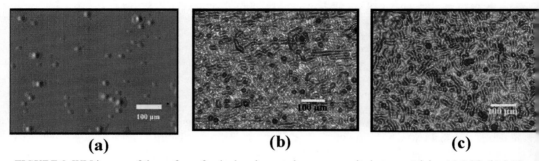

| (a) | (b) | (c) |

FIGURE 3. WLI images of the surface of a single spin coated nanocomposite layer containing (a) 0.05, (b) 0.15 and (c) 0.4 carbon MWNTs loading ratio.

Optical Properties

The optical transmission of the nanocomposite layers was investigated in the range between 400 and 3000 nm. Figure 5 shows the transmission of a 3 mm uncoated PC substrate and a spin coated layer of the conducting nanocomposite deposited on a PC substrate heated up to 130 °C for 30 min. The layer exhibits an average transmission higher than 85 % in the visible range (400 – 800 nm) and slightly lower than that of the PC substrate. The coating and the substrate show a strong absorption in the UV region (λ < 400 nm). The coatings on the glass substrate show similar behavior.

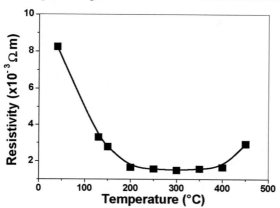

FIGURE 4. The electrical resistivity of a single spin coated carbon MWNTs composite layer with 0.4 loading ratio deposited on glass substrate as a function of the heating temperature.

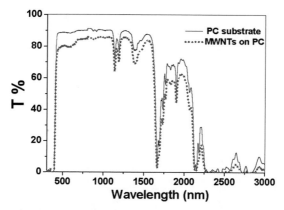

FIGURE 5. UV-VIS-NIR transmission curve for PC substrate and a spin coated carbon MWNTs composite layer deposited on PC substrate.

REFERENCES

1. B. G. Lewis and D. C. Paine, MRS Bulletin **25**, 22 (2000).
2. S. Chappelm and A. Zaban, Solar Energy Materials and Solar Cells **71**, 141 (2002).
3. R. Hatton, S. Day, M. Chesters and M. Willis, Thin Solid Films **394**, 292 (2001).
4. N. Al-Dahoudi and M. Aegerter, Surface Coatings International: Coating Transcations **88** (B4) 257-263 (2005).

5. X. Jing, W. Zhao and L. Lan, Journal of Materials Science Letters **19**, 377-379 (2000).
6. F. Lux, Journal of Materials Science **28**, 285-301 (1993).
7. M. Kaempgen, G. S. Duesberg, and S. Roth, Applied Surface Science **252**, 425-429 (2005).
8. Z. Wu, Z. Chen, X. Du, and J. M. Logan, J. Sippel, M. Nikolou, K. Kamaras, J. R. Reynolds, D. B. Tanner, A. F. Hebard, and A. G. Rinzler, Science **305**, 1273-1276 (2004).
9. J. Sun, W. Gerberich, and L. Francis, Journal of Polymer Science B **41**, 1744-1761 (2003).

Low-Dielectric Constant Polyimide Nanoporous Films: Synthesis and Properties

S. Mehdipour-Ataei, A. Rahimi and S. Saidi

Iran Polymer and Petrochemical Institute, P.O. Box 14965/115, Tehran, Iran
s.mehdipour@ippi.ac.ir

Abstract. Synthesis of high temperature polyimide foams with pore sizes in the nanometer range was developed. Foams were prepared by casting graft copolymers comprising a thermally stable block as the matrix and a thermally labile material as the dispersed phase. Polyimides derived from pyromellitic dianhydride with new diamines (4BAP and BAN) were used as the matrix material and functionalized poly(propylene glycol) oligomers were used as a thermally labile constituent. Upon thermal treatment the labile blocks were subsequently removed leaving pores with the size and shape of the original copolymer morphology. The polyimides and foamed polyimides were characterized by some conventional methods including FTIR, H-NMR, DSC, TGA, SEM, TEM, and dielectric constant.

Keywords: nanofoam; polyimide; phase separation; grafted copolymer
PACS: 77.84.jd

INTRODUCTION

The need for high temperature materials with low dielectric and thermal insulating behavior for industrial applications has increased in recent years [1]. As a class of materials, polyimides have been best satisfied the material requirements for these applications because they exhibit favorable balance of physical and chemical properties and show excellent thermal, mechanical and electrical properties [2]. However, their dielectric constants are not low enough to meet the specifications of intermetal dielectric layers. The most common approach in modifying the dielectric properties of polyimides has been focused on the incorporation of perfluoroalkyl groups. Examples include hexafluoroisopropylidine linkages and pendent trifluoromethyl groups [3].

The methodology for developing highly fluorinated polyimides can be limited to a certain extent by synthetic difficulties associated with the fluorine-containing comonomers. An alternative approach to lower polymer dielectric constant is to introduce nanoscopic porosity into the polymer film. The reduction in the dielectric constant could be achieved by replacing the polymer having a dielectric constant ~3.0 with air which has a dielectric constant of 1, while the desired thermal and mechanical properties were maintained. The size of the voids must be smaller than either the film thickness or any microelectronic features, i.e., much less than 1μm. These nanoporous (nanofoam) polymers can be prepared from block or graft copolymers comprised of a thermally stable and thermally labile materials, where the latter constitutes the dispersed phase [4]. Using a specified heat treatment, the thermally unstable block decomposes, leaving a porous structure of a size that is commensurate with the copolymer morphology, i.e., on the tens of nanometer size scale. The graft copolymer approach, alternative to block copolymer, where the thermally labile groups are attached to a polyimide backbone as a side chain, may be considered for improving the mechanical properties of the final nanofoam also in this system there is no limitation in growing of the molecular weight of polyimide backbone.

Here we describe preparation of benzophenone-based polyimide nanofoams grafted with poly (propylene glycol) and investigation of their physical and thermal properties.

CP929, *Nanotechnology and Its Applications, First Sharjah International Conference*
edited by Y. I. Salamin, N. M. Hamdan, H. Al-Awadhi, N. M. Jisrawi, and N. Tabet
© 2007 American Institute of Physics 978-0-7354-0439-7/07/$23.00

EXPERIMENTAL

Monomer and Polymer Synthesis

Synthesis of bis-[4-(4-amino-phenylene-1-yloxy)-phenyl]-methanone (4BAP) and bis-[4-(5-amino-naphthalene-1-yloxy)-phenyl]-methanone (BAN) diamines into a 100-mL, three-necked, round-bottomed flask equipped with a Dean-Stark trap, a condenser, a nitrogen inlet tube, a thermometer, an oil bath, and a magnetic stirrer was placed 0.01 mol of 4,4′-dichlorobenzophenone, 25 mL of dry N-methyl-2-pyrrolidone (NMP), and 15 mL of dry toluene and 0.021 mol of 4-amino phenol (for 4BAP diamine). Then 0.0315 mol of K_2CO_3 was added to the mixture and the reaction mixture was heated to 140 °C for 6h with continuous stirring. The generated water was removed from the reaction mixture by azeotropic distillation. The reaction temperature was raised to 165 °C by removing more toluene, and kept at the same temperature for 20h. During this time, progress of the reaction was monitored by thin-layer chromatography (TLC). The resulting reaction mixture was cooled and poured into water. Then 100 mL of 5% NaOH was added to the mixture and the mixture was washed repeatedly with a 5% NaOH solution and water. The obtained diamine was dried in a vacuum oven at 60 °C (Yield 90 %). The BAN diamine was prepared similarly in 88 % yield, using 5-amino-1-naphthol instead of 4-amino phenol.

RESULTS AND DISCUSSION

Aromatic diamines are valuable building blocks for the preparation of high performance polymers such as polyimides and inducing specific properties which make them suitable for certain applications. In this work we report the synthesis of three aromatic diamines 4BAP, 3BAP, and BAN and preparation of polyimide nanofoams therefrom. These diamines were prepared in one step by the aromatic nucleophilic substitution reaction of 4-amino phenol (for 4BAP diamine), and 5-amino-1-naphthol (for BAN diamine) with 4,4'-dichlorobenzophenone in the presence of anhydrous K_2CO_3 in NMP (Scheme 1).

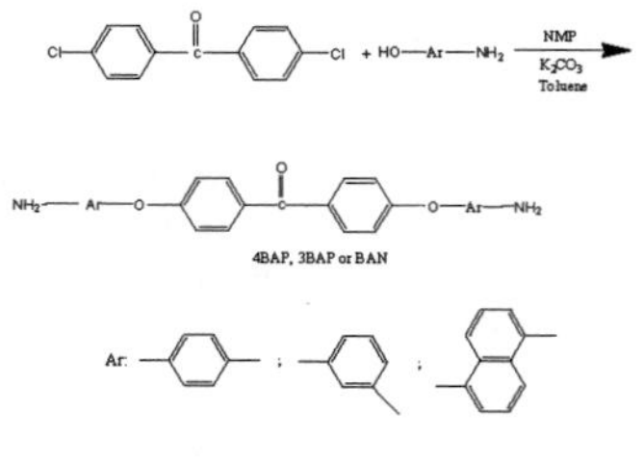

SCHEME 1. Structure of diamines.

Structures of the prepared diamines were characterized by spectroscopic methods. The prepared diamines were used for the synthesis of polyimides and subsequently polyimide nanofoams. Presence of flexible ether and carbonyl groups would induce flexibility to the final polyimide films and also presence of aromatic groups and avoidance of any weak linkages in the structure of diamines would maintain thermal stability of the polymers. By preparing these diamines the effect of substitution (meta and para catenation) and bulkiness of the naphthyl group on the properties of the final polyimides could be studied. The graft copolymer approach used to generate polyimide nanofoams requires well-defined block segments, which include thermally stable matrix block (imide units) and the thermally labile block (PPG units), the latter was the disperse phase. The phase-separated domain of the labile block was thermally decomposed, leaving a size and shape corresponding to the initial morphology. The temperature at which decomposition occurs was crucial. The decomposition temperature must be sufficiently high to permit standard film preparation and solvent removal yet well below the Tg of the polyimide block, to avoid foam collapse [9]. One criterion for the thermally decomposable block is the preparation of well-defined functional oligomers. Thus, poly (propylene glycol) monobutyl ether (PPGOH) with molecular weight of 1000 and 2500 g mol^{-1} were functionalized by reaction with 2-bromo acetylbromide in the presence of triethylamine and used as labile blocks (Scheme 2).

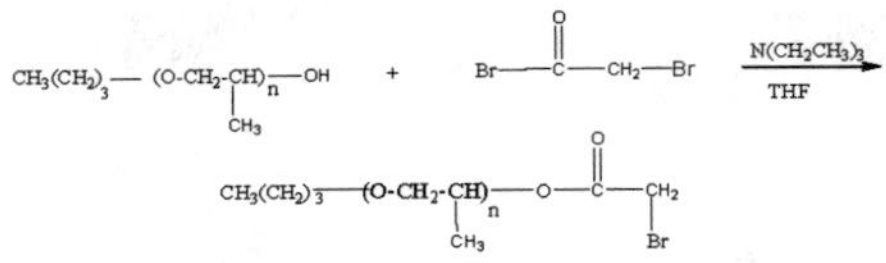

SCHEME 2. Structure of PPG-Br.

The graft copolymers (PAAE-g-PPGs) were prepared via poly (amic acid) precursor. Solution polycondensation of pyromellitic dianhydride with 4BAP and BAN diamines led to preparation of related poly (amic acid)s. Then PPG-Br oligomers were added to the poly (amic acid) solutions in the presence of K_2CO_3. Under these conditions, nucleophilic substitution of carboxylate anion onto the bromide-substituted carbon occurred. Then the viscous and brown solutions were precipitated in methanol/water (1:1 v/v) to eliminate unreacted PPG-Br (Scheme 3).

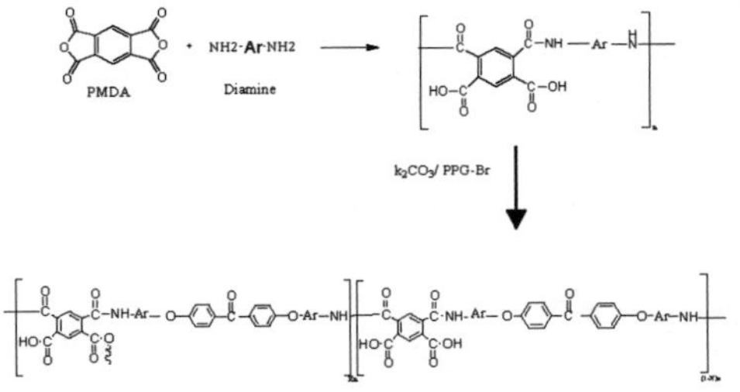

SCHEME 3. Synthesis of PAAE-g-PPG.

After solution casting of the polymers on the glass substrates, the samples were subjected in two thermal cycles: At first, samples were heated from 40 °C to 180 °C at a heating rate of 20 °C/hr for 7 h under nitrogen atmosphere to remove solvent and to do imidization process. The formation of polyimides was confirmed by IR spectroscopy. Then they were followed by heating at 300 °C for 9 h in air to complete decomposition of PPG moieties and foam formation.

FT-IR spectroscopy and TGA analysis were performed to monitor thermolysis of PPG from copolymer samples. No detectable residual labile block or PPG byproducts were observed in the resulting porous polymers after foam formation. In the FT-IR spectra of the prepared foams the peaks at about 2927 cm^{-1} related to the aliphatic group of PPG were disappeared (Figure 1) and no weight losses were detected at about 320 °C for the samples in their TGA curves due to the decomposition of PPG block.

Thermal behavior and thermal stability of the prepared poly (amic acid)s and related polyimides were studied using DSC and TGA methods. Thermal stability of the polyimides was increased according to the following trends of diamines: 4BAP > BAN, which was attributed to the symmetry of the diamines and better growth of molecular weights. Also, the polyimides with same diamine structures and lower molecular weights of polyols were more thermally stable than the higher molecular weights of polyols which was related to the lower amounts of aliphatic weak linkages in the structure. However, polyimide nanofoams showed a reverse trend of thermal stability probably due to the more decomposition of lower molecular weights of polyols in comparison with higher molecular weights of polyols in foaming thermal process. The results of thermal stability for polyimide nanofoams were collected in Table 1.

It should be mention that, a three-step decomposition behavior was seen in the TGA thermographs of the poly (am acid)s. The first mass loss at about 100 °C was related to the H$_2$O elimination due to the thermal imidization of the am acids. The second mass loss at about 320 °C was attributed to the degradation of PPG units and the third mass loss about 520 °C was related to decomposition of the main chain of the polyimides.

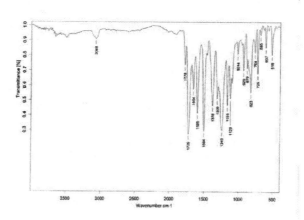

FIGURE 1. FT-IR spectrum of 4BAP-2500 based foam.

TABLE 1. Thermal properties of the polyimides nanofoams.

Polymer	T$_0$(°C)	T$_{10}$(°C)	T$_{max}$(°C)	Char yield at 600°C(%)
4BAP-1000	410	525	560	56
BAN-1000	370	492	547	49
4BAP-2500	425	535	565	58
BAN-2500	380	500	555	52

T$_0$: initial decomposition temperature, T$_{10}$: temperature for 10% weight loss, T$_{max}$: maximum decomposition temperature, Char Yield: weight polymer remained

The dielectric constants for original polyimides and nanoporous films were measured with an optical method an calculated via Maxwell's identity ($\varepsilon = \eta^2$, $\Delta \varepsilon = + 0.3$) [5] and the results are collected in Table 2. According to th obtained results, the difference between the in-plane and out of plane refractive indices is in the range of 0.03-0.0 which represents a good estimation of the anisotropy. The refractive indices and therefore dielectric constants of th nanofoams are smaller than the related polyimides due to the incorporation of pores into the final structures.

TABLE 2. Refractive indices and dielectric constants of the polyimides and their nanofoams.

Polymer[1]	η_i[2]	η_o[3]	$\Delta\eta$	ε_o[4]
4BAP (h)	1.74	1.71	0.03	2.92
4BAP-1000 (n)	1.70	1.66	0.04	2.75
4BAP-2500 (n)	1.68	1.64	0.04	2.68
BAN (h)	1.70	1.65	0.05	2.72
BAN-1000 (n)	1.66	1.60	0.06	2.56
BAN-2500 (n)	1.64	1.57	0.07	2.47

[1]h: homopolymer; n: nanopolymer; [2]η_i: in-plane refractive index; [3]η_o: out-of-plane refractive index; [4]ε_o: out-of-plane dielectric constant.

According to the SEM and TEM micrograph, a nanophase-separated morphology was observed for the polyimide nanofoams. A uniform distribution of pore units and little interconnectivity between the pores were concluded from the micrographs. The average pore size of the polyimide nanofoams was found to be in the range of 10 – 20 nm (see Figs. 2 and 3).

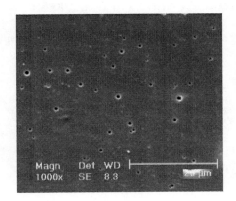

FIGURE 2. SEM micrograph of BAN-2500 based foam.

FIGURE 3. TEM micrograph of BAN-2500 based foam

CONCLUSION

In order to study the structure-property relations of thermally stable matrix block also for investigation and effect of thermally labile block on the final properties of the polyimide nanofoams three benzophenon-based diamines and two different PPG were used for the synthesis of these nanofoams. Thus, six different polyimide nanofoams with high thermal stability and low dielectric constant were obtained via grafting of PPG on polyimide backbones. They can be expected to be useful for the practical applications in thermally stable insulating materials.

ACKNOWLEDGMENTS

The authors would like to thank the Iranian Nanotechnology Initiative (INI) for partial support of this research.

REFERENCES

1. P. E. Cassidy, *Thermally Stable Polymers* (Marcel Dekker, New York, 1980).
2. S. Mehdipour-Ataei, N. Bahri, and A. Amirshaghaghi, *Polym. Degrad. Stabil.* **91**, 2622 (2006).
3. D. Wilson, H. D. Stenzenberger, and P. M. Hergenrother (*Polyimides*, Blackie, London, 1990).
4. P. M. Hergenrother, K.A.Watson, J. G. Smith, and J. W. Connell, *Polymer* **43**, 5077(2002).
5. D. Boese, H. Lee, D. Y. Yoon, J. D. Swalen, and J. F. Rabolt, *J. Polym. Sci. B: Polym. Phys.* **30**, 1321(1992).

Effects of Induction and Convention Aging on Growth Kinetics and Distribution of Nanometric γ' Precipitates in an Ni-Based Superalloy

A. Abdollah-Zadeh, S. Soleimani, A. Samadi and H. Assadi

Department of Materials Engineering, Tarbiat Modares University, P. O. Box: 14115-143, Tehran, Iran
zadeh@modares.ac.ir

Abstract. In this work, the effect of induction aging on the microstructure and the hardness of a cast superalloy, Udimet500, is investigated. The solution samples were aged in two different furnaces, namely, induction furnace and resistance tube furnace, at 850°C up to 2 hours. The age-hardening behavior and microstructural characteristics were studied by hardness testing, scanning electron microscopy (SEM), electron image analyzing, X-ray diffractometery, electrolytic extraction and light laser scattering (LLS). The results show differences in the distribution of precipitates in the two groups of samples, while showing formation of tertiary fine precipitates in both groups. Formation of these fine precipitates appears to have an influence on the mechanical properties of the material.

Keywords: Induction, Aging, Superalloy, Udimet500, Growth, Nucleation.
PACS: 00

INTRODUCTION

The mechanical properties of an Ni-based superalloy are dependent upon such factors as volume fraction, particle size, nucleation, growth and coarsening rates and composition of γ' phase [1-3]. Moreover, studies show that the nucleation and growth of γ' precipitates in Ni-based superalloys is diffusion controlled [4-6]. Therefore, the rate of nucleation and growth is dependent upon the time and temperature of the process [1].

Work by Razavi and coworkers [7] showed that induction aging on In738LC leads to the higher rate of nucleation and growth in comparison with conventional resistance and salt bath aging in similar time and temperature. That observation was the basis of the present work and a different kind of Ni-based superalloy was chosen to apply the induction aging on it. The aim of this study is to investigate the increasing effect of induction on the rate of nucleation and growth processes in the superalloy samples. For this reason, two types of age hardening, namely induction aging and resistance tube furnace aging were conducted on a Ni-based cast superalloy, Udimet 500, to investigate the effect of induction on this alloy. Moreover, the effect of induction aging time on the microstructure is investigated for better understanding the microstructural evolution of the alloy.

EXPERIMENTAL PROCEDURE

The cast alloy used in this study was Udimet 500 with the chemical composition (in wt%) of 19.1Cr, 16.20Co, 3.70Al, 3.05Ti, 4.20Mo, 0.40Fe, 0.02C, 0.02Si and balance nickel. Cylindrical samples with the height and the diameter of 20 and 8 mm, respectively, were machined. A hole was made on one side of each sample with the depth and inner diameter of 5 and 1 mm, respectively. This hole was used for controlling the temperature of the sample by putting the

CP929, *Nanotechnology and Its Applications, First Sharjah International Conference*
edited by Y. I. Salamin, N. M. Hamdan, H. Al-Awadhi, N. M. Jisrawi, and N. Tabet
© 2007 American Institute of Physics 978-0-7354-0439-7/07/$23.00

thermocouple of the induction furnace into it. The temperature and the quench rate of the solution treatment were investigated by the hardness and the size distribution of precipitates. Samples were solution treated in the selected condition at 1180°C for 4 hours and quenched in iced-Brine. In order to investigate the effect of induction on the aging process and the rate of diffusion, samples were aged in two different furnaces, namely induction and conventional resistance tube furnace. In each furnace, two samples were aged at 850°C for 60 and 120 min followed by iced-Brine quenching. The heating rate in both furnaces was 141°C/min. The atmosphere of both furnaces for all the tests was air.

The effect of induction aging time was investigated by aging the samples at 850 °C for 15, 30, 45, 60 and 120 min in induction furnace followed by quenching in iced-brine. Temperature and the heating rate of the samples in induction furnace were programmed and controlled by an electronic programmable device which was connected precisely to the induction furnace and thermocouple.

The hardness of the samples was measured by using the HV method. The samples were electro-etched in 5% H_2CrO_4 for γ' observation. Microstructures were examined by scanning electron microscope (SEM) and the γ' precipitates were characterized by image analyzer. In this stage, the size and volume fraction of γ' precipitates were measured in all the samples. The weight percent of γ' precipitates of the samples were measured using electrolytic extraction. Samples were immersed in 20% H_3PO_4 at room temperature for 1 hour and the obtained solutions were filtered. In order to extract and weigh of very fine γ' precipitates, the filtered solutions were centrifuged. The precipitates of the filters were investigated by X-ray diffractometery to assure the absence of other kinds of precipitates. Furthermore, the electrolytic solutions of the samples were investigated by Light Laser Scattering (LLS) technique to obtain the size distribution of the γ precipitates. All the parameters of each solution, such as viscosity, were measured and set on the device to obtain the precise results.

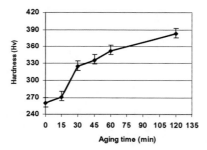

FIGURE 1. Variation in hardness with time for induction aging temperature of 850°C.

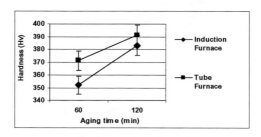

FIGURE 2. Variation in hardness with time for two furnaces at aging temperature of 850°C.

112

RESULTS AND DISCUSSION

The hardness versus time curve for induction samples at the aging temperature of 850 °C is shown in Fig. 1. It can be seen that the hardness of induction samples is increasing with the increasing the aging time up to 2 hours. It implies that the nucleation and growth of γ' precipitates are the predominant processes during 2 hours aging without any evidence of coarsening.

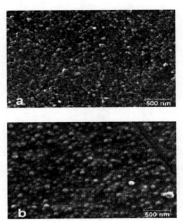

FIGURE 3. SEM images of the samples aged at 850°C for 2 hours in a) resistance tube furnace, and b) induction furnace.

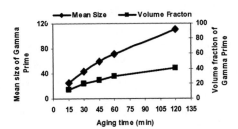

FIGURE 4. The mean size and volume fraction of γ' precipitates versus aging time for induction samples.]

Figure 2 illustrates a comparison between the hardness results of induction and conventional tube furnaces. It indicates that the amounts of hardness of induction samples are lower than the samples aged at similar conditions in tube furnace. This result can be related to two factors: a) the wt% of γ' precipitates, and b) the size distribution of γ' precipitates. These two factors will be discussed in the following sections.

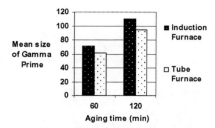

FIGURE 5. The mean size of γ' precipitates versus aging time for the samples aged at two different furnaces.

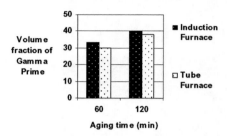

FIGURE 6. The volume fraction of γ' precipitates versus aging time for the samples aged at two different furnaces.

The typical SEM micrographs of samples after aging in two different furnaces are shown in Fig. 3. Figure 4 presents the mean size and volume fraction of γ' precipitates versus induction aging time. Two bar charts for comparing the volume fractions and mean sizes of γ' precipitates resulted by aging in two furnaces are also illustrated in Figs. 5 and 6, respectively.

Analysis of microstructure of the induction samples shows that γ' precipitates are fine, dispersed and almost spherical shaped and are randomly distributed in the microstructure. Two types of γ' precipitates in aging times more than 45 min were identified: a) the larger precipitates with more than 100 nm diameter as the secondary γ', and b) the fine precipitates with less than 100 nm diameter as the tertiary γ'.

Study of the results of electron image analysis indicates that in the induction aging process, the increase of time at 850 °C results in the increase in the size and the volume fraction of γ' precipitates.

As can be seen in Fig. 6, the volume fraction of γ' precipitates in induction samples are more than that of tube furnace samples. It is initially expected from this result a higher amount of hardness in induction samples. However, this statement is in contradiction to the hardness results shown in Fig. 2. Consequently, the amount of volume fraction of precipitates cannot be the predominant factor. In this case, the size distribution of γ' precipitates can explain the hardness results observed in Fig. 2.

The filtration and centrifuging processes were done on the suspensions made by electrolytic extraction and the weight fractions of γ' precipitates were measured. A very good agreement was observed between the volume fractions resulted by electron image analyzer and weight fractions obtained.

The electrolytic suspensions were investigated by the LLS analyzer to obtain the size distribution of γ' precipitates. The size distribution curves for different induction aging times are shown in Fig. 7. To compare the size distributions of γ' precipitates for two furnaces, the related curves are illustrated in Fig. 8.

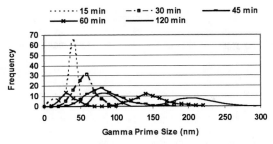

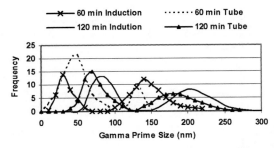

FIGURE 7. Particle size histograms for different induction aging times.

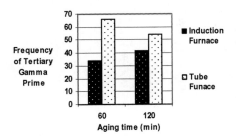

FIGURE 8. Particle size histograms for the samples aged at two different furnaces.

As can be seen in Fig. 7, with the increasing the aging time, the size of γ' precipitates increases and in contradiction to the width of peaks, the height of the peaks decrease to keep the surrounded area constant.

FIGURE 9. Particle size bar chart of tertiary γ' precipitates for the samples aged at two furnaces.

Fig. 8 shows that the size distribution of γ' precipitates for induction samples is larger than that of tube furnace samples. Moreover, a new peak with the mean diameter less than 100 nm in the samples of both groups can be observed. For induction samples, these new peaks can be the result of nucleation of tertiary γ' precipitates after 45 min aging.

115

The weight percents (wt%) of tertiary γ' precipitates for the samples of two furnaces are shown in Fig. 9. This result show that the wt% of tertiary γ' precipitates for induction samples is lower than that of tube furnace samples. On the other hand, the mean size of tertiary γ' precipitates for induction samples is higher than that related to the tube furnace samples.

Although the wt% of γ' precipitates in tube furnace samples is lower than that of induction samples, but it can be concluded that the presence of finer precipitates in tube furnace samples has led to the higher hardness.

Fine γ' precipitates form due to a sharp decrease in the rate of growth and increase in the rate of nucleation. At the beginning of the aging process, particles of secondary γ' nucleate in supersaturated matrix of γ. With the progress of aging, growth of secondary γ' occurs with the partitioning the solute into the surrounding matrix, possibly because the γ' precipitates cannot grow fast enough to decrease the supersaturation of the matrix. Eventually, at a critical aging time, the driving force for the nucleation is sufficient to allow nucleation of tertiary γ'.

CONCLUSIONS

In the present study, two types of heating, namely Induction aging and resistance tube furnace aging were conducted on a chosen Ni-based cast superalloy, Udimet500 and the results were analyzed and compared. The solution samples were aged in both furnaces up to 2 hours. The age-hardening behavior and microstructural characteristics were studied by hardness testing, scanning electron microscopy (SEM), electron image analyzing, X-ray diffractometery, electrolytic extraction and light laser scattering (LLS).

The results are summarized as follows:

- Induction aging of Udimet 500 may lead to the formation of nano-sized γ' precipitates.
- LLS analyses of samples aged to different times show a bi-modal distribution of precipitates for aging more than 1 hour.
- After induction aging at 850°C for 2 hours, nucleation and growth of γ' precipitates take place without any evidence of coarsening process.
- The rate of nucleation and growth in induction aging is higher than that of conventional resistance furnace aging. This observation is in agreement with previous results for In738LC Ni-based superalloy.

REFERENCES

1. C. T. Sims, N. S. Stoloff, and W. C. Hagel, Superalloy II (Wiley, New York, 1987) pp. 97-131.
2. A. Baladan, Mater. Sci. **37**, 2379-2405 (2002).
3. M. P. Jackson and R. C. Reed, Mater. Sci. Eng. A **259**, 85-91 (1999).
4. A. Ges, H. Palacio and R. Versaci, Mater. Sci. **29**, 3572-3580 (1994).
5. T. Grosdidier, A. Hazotte and A. Simon, Mater. Sci. Eng. A **256**, 183-188 (1998).
6. M. F. Henry, Y. S. Yod, D. Y. Yoon and J. Choi, Metal. Trans. A **24A**, 1733-1740 (1993).
7. S. H. Razavi, Sh. Mirdamadi, J. Szpunar, and H. Arabi, Mater. Sci **37**, 1461-1469 (2002).

Role of Particle Size and Crystal Microstructure on the Magnetic Behaviour of Binary Alloy Nanoparticles

A. S. Hussein, P. Murugaraj, C. J. Rix and D. E. Mainwaring

School of Applied Sciences, Royal Melbourne Institute of Technology, Melbourne, Victoria 3000
david.mainwaring@rmit.edu.au

Abstract. The increasing research attention towards magnetic nanoparticles has been driven by interest in advanced device and biomolecular technologies. Particle size and crystal microstructure represent key structural aspects in achieving optimum performance of nanoparticle systems. To date these relationships have been studied in sputtered alloy nanoparticle thin films. In this paper, we report the relationships between the thermal annealing parameters and the physical and magnetic characteristics of CoPt alloy nanoparticles.

Keywords: Chemical synthesis, reverse micelles, metal nanoparticles, crystal structure, magnetic.
PACS: 61.46.−w

INTRODUCTION

There is a growing interest in the synthesis and characterizations of nanoscale magnetic materials due to the novel magnetic properties they exhibit. Chemically synthesised CoPt nanoparticles have attracted much attention due to their narrow particle size distribution at the nanometer scale compared to those fabricated by existing sputtering techniques. Applications of such materials may include magnetic data storage media [1], highly sensitive magnetic bioseparation [2], magnetic biosensing [3] and advanced permanent nanocomposite magnets [4] where chemical synthesis provides good control of nanoparticle size. Borohydride reduction of metal salts in reverse micelles was first pioneered by Pileni *et al* [5]. The as-synthesized CoPt nanoparticles exhibited a disordered face centered cubic (fcc) crystal structure with superparamagnetic character. Relationship between crystal structure and magnetic properties with annealing temperature and duration have been reported for magnetic thin films and as well as magnetic nanoparticulates prepared by different chemical [6] and physical techniques [7]. In this work, we report on the control of magnetic properties through annealing induced particle size, morphology and crystal structure in CoPt alloy nanoparticles by a synthesis method in which the actual reduction of cobalt and platinum salts and the subsequent formation of the CoPt nanoalloy occurs within the aqueous core of a reverse micelle.

EXPERIMENTAL SECTION

The synthesis of CoPt nanoalloys was performed according to a slightly modified method first used by Pileni et al. [5]. A NaAOT reverse micellar solution of cobalt chloride and sodium tetrachloroplatinate at the desired metal concentration and water to surfactant ratio, $w = [H_2O]/[NaAOT] = 10$, was prepared in heptane. Another reverse micellar solution of sodium borohydride with same w=10 was prepared separately. These solutions were mixed such that the borohydride was 150% stoichiometric excess to metal ions. The resulting solution turned from golden-brown to black, indicating the formation of metallic nanoparticles which were recovered from the micelles with addition of

CP929, *Nanotechnology and Its Applications, First Sharjah International Conference*
edited by Y. I. Salamin, N. M. Hamdan, H. Al-Awadhi, N. M. Jisrawi, and N. Tabet
© 2007 American Institute of Physics 978-0-7354-0439-7/07/$23.00

ethanol. These nanoparticles were then extracted, washed and annealed under nitrogen at temperatures ranging from 60°C to 600°C.

The magnetic properties of all samples were determined using a vibrating sample magnetometer (VSM) (Molspin "Nuvo") operating at a maximum field of +/- 1.0 Tesla (-/+ 10 kOe) and as well with a SQUID magnetometer (Quantum Design Magnetic Property Measurement System) operating at +/- 7.0 Tesla. An X-ray diffractometer (Bruker D8 Advance) equipped with a CuK$_\alpha$ radiation source was used to obtain the diffraction patterns of all nanoparticle powders. All X-ray data was obtained in the θ-2θ locked-couple mode over a 2θ interval of 20-90 degrees. A high-resolution transmission electron microscope (JEOL 2010) operating at 200KV was used to determine microstructural characteristics of the nanocrystallites.

RESULTS AND DISCUSSION

Structural Investigation of CoPt alloys

The synthesized CoPt, extracted, purified and dried at 60°C, was a nanocrystalline material confirmed by the XRD measurements as shown in Fig. 1a, which could be indexed to a face centered cubic (fcc) structure with a lattice constant of 3.8567 ± 0.0003 Å (A1 phase). In this phase, the Co and Pt have anti-site disorder. The average crystallite size of the CoPt nanoparticles calculated using the Scherrer formula [8] was ~ 4 nm. The dried nanocrystalline powder was then subjected to annealing treatments to 550°C for one hour duration. XRD of these annealed samples showed that the A1 phase is stable to 550°C (Fig, 1a).

When the CoPt nanocrystalline sample was heated at 600°C for one hour, there was a significant change in the XRD pattern, with additional peaks together with shifts in the existing peaks (see Fig. 1a). This new structure can be indexed to the well known fct phase of the CoPt also known as L1$_0$ phase. This phase transformation occurs due to the ordering of the Co and the Pt ions in their respective sublattices. The ordering extends the CoPt unit cell along the a$_0$ axis and contracts along c$_0$-axis. The lattice constants calculated for this sample were, a$_0$ = 3.756 ± 0.003 Å, and c$_0$ =3.711 ± 0.003 Å, with an axial ratio (c$_0$/a$_0$) of 0.9856 ± 0.0001. The axial ratio is proportional to the degree of tetragonal distortion (1.44 %) and the extent of ordering of the Co and Pt ions in their respective sublattices. Bulk CoPt with a fct crystal structure has lattice constants of a$_0$ = 3.803 ± 0.003 Å and c$_0$ =3.701 ± 0.003 Å giving an axial ratio c$_0$/a$_0$ of 0.973 ± 0.003 Å, corresponding to a tetragonal distortion of 2.7 %. This shows that the samples annealed at 600 °C for one hour were only partially ordered and the remaining phase still exists as a disordered cubic phase. This is in good agreement with reported CoPt nanoparticles [9].

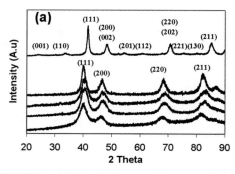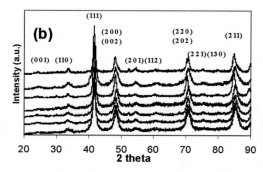

FIGURE 1. (a) XRD of CoPt nanoparticles synthesized at room temperature and dried for 60 °C, and annealed for 1 h. at 350°C 450°C 550°C 600°C; (b) Changes in the XRD patterns of CoPt annealed at 600°C for increased annealing times 0.5, 1, 2, 4, 6, 8 and 10 h. (plots from bottom upwards)

Figure 1b shows XRD patterns obtained with samples annealed at 600 °C for times varying from 30 minutes to 10 hours. With increased time of annealing at 600 °C, the degree of ordering increased as observed by the continuous decrease in c_o/a_o (Fig. 2a) When the long range ordering is established in the Co and Pt sublattice, a CoPt supercell is formed as indicated by the appearance of superstructure peaks in the XRD pattern (Fig. 1b) that can be indexed to (001), 110), (201), (112), (221) and (130) lattice planes.

The degree of ordering may be described by the long-range order parameter S given by [10]

$$S = 2\left(r_{Co} - x_{Co}\right) = 2\left(r_{Pt} - x_{Pt}\right),\tag{1}$$

where x_{Co} and x_{Pt} are the atomic fractions of Co and Pt in the alloy, respectively, and r_{Co} and r_{Pt} are the fractions of Co sites or Pt sites, respectively, occupied by the correct atom. If all the sites are occupied by the correct atoms then $r_{Co} = r_{Pt} = 1$. From this expression, for the equiatomic CoPt alloy, S values were derived ranging from 0 to 1 which correspond to a disordered state and a perfectly ordered lattice, respectively, with intermediate values representing partial long-range order.

The theoretically evaluated S is directly proportional to the integrated intensity of the X-ray superstructure reflections. Experimentally, S was evaluated by comparing the intensity of superstructure reflections with those of the fundamental reflections from each individual XRD pattern. The order parameter increased with time of annealing at 600°C indicative of increased ordering in the Co and Pt sublattices (Fig. 2b). This result is also consistent with the increased tetragonal distortion with annealing time as shown in Fig. 2a. The calculated order parameter 'S' increased from about 0.04 to 0.1 for samples annealed for 10 hours at 600°C while the tetragonal distortion increased from 1.43% to 6.63%

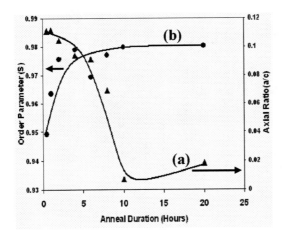

FIGURE 2. (a) Variation of the axial ratio and (b) Calculated order parameter "S" obtained from XRD data of fct CoPt nanoparticles with increasing annealing time at 600°C (continuous line is a visual guide only.).

The CoPt crystallite size as obtained from the XRD data using Scherrer's formula increased with annealing temperature (Fig. 1a). The crystallite size annealed at temperatures up to 550°C (1h.) gradually increased to 6nm and rose to 13nm at 600°C (1h). Annealing at 600°C for 10 h yields a crystallite size of 17nm. The formation of the fcc phase at low temperatures and the subsequent formation of the fct phase is consistent with the reports of both the CoPt and FePt based thin films [11] and nanoparticles [12].

HRTEM images of the as synthesized CoPt nanoparticles showed that the particles have spherical morphology with a diameter of 4-5 nm as shown in Fig. 3(a). Selected Area Diffraction patterns obtained on these crystallites (Fig. 3b) were be indexed to a fcc structure consistent with the results obtained from powder XRD studies. The bright field

HRTEM image of a single nanoparticle annealed at 600°C for one hour is shown in Fig. 4a. Figure 4b shows the Selected Area Diffraction pattern obtained on this crystallite which was indexed to the L1$_0$ phase of the CoPt alloy.

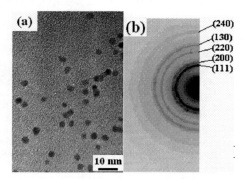

FIGURE 3. (a) Bright field HRTEM image of the as-synthesized CoPt nanoparticles, and (b) the corresponding indexed electron diffraction pattern.

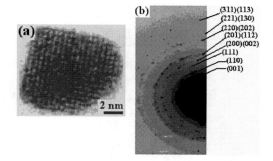

FIGURE 4. (a) HRTEM image of a single CoPt nanoparticle annealed at 600°C for 1 h. (b) the corresponding indexed electron diffraction pattern.

Magnetic Property Investigations of CoPt Nanoparticles

The CoPt nanocrystalline samples synthesised at room temperature and dried at 60°C exhibited superparamagnetic behavior with a very low coercivity as shown in Fig. 5, which arises from the grain size being below the magnetic domain size as well as the disordered cubic structure. These nanocrystallites do not have any magnetic remenance (M_r) but possess very high saturation magnetization (M_s) as shown in plot (a) of Fig. 5 which is typical of superparamagnetic behaviour. Annealing at temperatures up to 550°C did not appreciably enhance the ferromagnetic properties of the CoPt nanoparticles (plot (b) of Fig.5). Such magnetic behaviour is consistent with XRD structural analysis which showed that the CoPt nanocrystallites retained a disordered fcc lattice at these annealing temperatures. Hence, the CoPt samples annealed at 550°C for 1 hour may be classified as weakly ferromagnetic with an H_c value < 350 Oe, a M_s value of 2.5 emu/g (39.9 emu/cc) and a M_r value of 0.8 emu/g (12.8 emu/cc) as shown in Fig. 5. The CoPt nanocrystallites heat treated at 600°C for 30 mins and possessing fct crystal structure exhibited considerable enhancement in the H_c (six-fold increase to a value ~2000 Oe), M_r values of 5.1 emu/g (81.4 emu/cc) and saturation magnetization (M_s) of 8.8 emu/g (140.4 emu/cc) (plot (c) of Fig. 5). This large difference is consistent with the structural data obtained from XRD measurements which show the transformation from a disordered A1 (fcc) phase to an ordered L1$_0$ (fct) phase (Fig. 1).

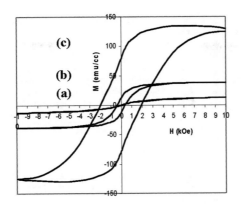

FIGURE 5. The magnetic hysteresis of CoPt nanoparticles (a) as-synthesised; (b) after heat treatment at 550°C for 1h.; (c) after heat treatment at 600 °C for 1h.

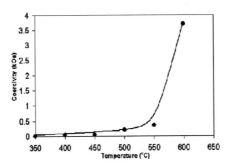

FIGURE 6. Variation of magnetic coercivity of CoPt samples annealed at different temperatures between 350 °C -600°C for 1 h.

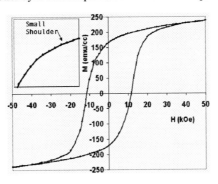

FIGURE 7. Magnetic hysteresis curve of CoPt alloy nanoparticles annealed at 600 °C for 10 h. The inset is an enlargement of the second quadrant of the hysteresis curve.

121

Figure 6 shows the variation in magnetic coercivity as a function of annealing temperature clearly indicating the large enhancement between 550°C and 600°C.

Figure 7 provides the hysteresis loop measured for CoPt nanocrystalline sample heat treated at 600°C for 10 hours. The H_c value of 12kOe was obtained while it is also exhibited a broad 'shoulder' indicative of a mixture of hard and soft magnetic phases, which in the present is case indicative of existence of ordered fct phase together with a disordered fcc phase. This H_c value appears to be highest reported to date for CoPt alloy nanoparticles, being more than 35 times larger than the value observed for the as-synthesized alloy. The saturation magnetization calculated from magnetization data was 15.5 emu/g (247.4 emu/cc) with a remanent magnetization of 11.0 emu/g (175.6 emu/cc).

REFERENCES

1. D. Weller, A. Moser, L. Folks, M. E. Best, W.Le, M. F. Toney, M. Schweicker, J.-L. Thiele and M. F. Doerner. IEEE Trans. Magn. **36**, 10-14 (2000).
2. Y. D. Ivanov, V. M. Govorun,V. A. Bykov and A. I. Archakov, *Proteomics*. 6, 1399-1414, (2005).
3. Z. Jiang, J. Llandro, T. Mitrelias and J. A. C. Bland, J. Appl. Phys. 99, 08S105/1-08S105/3. (2006).
4. M.Yue, P. L.Niu, J. X. Zhang, D. T.Zhang, IEEE Trans on Magn. **42**, 2894-2896, (2006).
5. M. P. Pileni, J. Phys. Chem. B 97, 6961-6973 (1993)
6. S. Sun, C. B. Murray, D. Weller, L. Folks, A. Moser, Science 287, 1989-1991 (2000).
7. H. Wang, S. X. Xue, F. J. Yang, H. B. Wang, X. Cao, J. A. Wang, Y. Gao, Z. B. Huang, C. P. Yang, W. Y. Cheung, S. P. Wong, and Q. Li,Z. Y. Li, Thin Solid Films **505**, 77-80 (2006).
8. B. D. Cullity, Elements of X-Ray Diffraction (Adison_Wesley Publishing, London, 1978) pp. 28-29.
9. R. A. Ristau, K. Barmak, L. H. Lewis, K. R. Coffey and J. K. Howard, J. Appl. Phys. **86**, 4527-4533 (1999).
10. B. E. Warren, X-ray diffraction (Dover, New York, 1990) p. 206-211.
11. S. Jeong, Y-N. Hsu, D. E. Laughlin, and M. E. McHenry, IEEE Trans. Magn. **36**, 2336-2338. (2000).
12. X. Sun, Y. Huang and D. E. Nikles, Intern. J. Nanotech. **1**, 328-346, (2004).

Metal – Insulator Transition-like in Nano-Crystallized Ni-Fe-Zr Metallic Glasses

F. Hamed, I. M. Obaidat and M. Benkraouda

Department of Physics, United Arab Emirates University, P.O. Box 17551 AlAin, UAE,
FHamed@uaeu.ac.ae

Abstract. Ni-Fe-Zr based Metallic glassy ribbons were prepared by melt spinning technique. The compositional and structural integrity of the melt spun ribbons were verified by means of X-ray diffraction, SEM, EDX and DSC. 5 to 7 cm long ribbons of Ni-Fe-Zr based metallic glasses with different compositions were sealed inside quartz ampoules under vacuum. The sealed metallic glassy ribbons were nano-crystallized at 973 K for varying periods of time. The temperature dependence of the electrical resistivity of the nano-crystallized samples had been investigated over the temperature range 25-280 K. The crystallized ribbons at 973 K for periods for less than 4 hours displayed insulating electrical behavior like at low temperatures, while those annealed for more than 4 hours showed metallic behavior like. Nonlinear I-V characteristics were also observed at low temperatures for samples annealed for less than four hours.

Keywords: Metal-Insulator Transition, Nano-crystallization of Metallic Glasses, Metallic Glasses.
PACS: 72.15.Cz, 71.23.Cq, 61.43.Dq, 81.05.kf, 81.07.Bc

INTRODUCTION

Metallic glasses have become the subject of interest to a great number of scientists and technologists since they were first discovered in 1960 [1]. Mizutani has classified metallic glasses into five groups based on their electronic properties [2]. Non-magnetic metallic glasses were classified as groups 4 and 5. Group 4 metallic glasses are defined as those for which the Fermi level is in the d band, while the Fermi level for group 5 is in the sp band. The temperature dependence of the electrical resistivity ($\rho(T)$) of group 5 metallic glasses over the temperature range 30-300 K can be understood within the framework of the generalized Faber-Ziman theory [2, 3]. While for group 4 metallic glasses over the range 30-300 K, Mizutani proposed the following empirical equation $\rho(T)/\rho(300 \text{ K}) = A + B \exp(-T/\Delta)$, where A, B and Δ are fitting parameters [3]. At temperatures below 30 K, superconductivity and quantum interference effects arise [4-7].

Nano-crystallization of metallic glasses is an active field that promises materials with excellent physical properties [8-10]. Saida *et al.* have found that icosahedral quasicrystalline particles ranging from 5 to 10 nm in diameter have precipitated in $Zr_{70}Fe_{20}Ni_{10}$ ternary metallic glass when it was annealed at 670 K [11]. Altounian *et al.* have found that the crystallization of $FeZr_2$ metallic glass proceeds from the evolution of a large number of nano-crystallites 2.0-3.0 nm in size and further annealing at 900 K for 2 hours resulted in grain growth of about 1 µm in size [12]. Liu *et al* have found recently that the crystallization of $Fe_{33}Zr_{67}$ and $Fe_{10}Zr_{90}$ is temperature and time dependent [13]. Dikeakos *et al* have found insulating electrical behavior at low temperatures in crystallized $FeZr_2$ metallic glasses; however, there was no mention of any changes in this insulating electrical behavior as a function of annealing temperature and time [14]. The purpose of the present study is to investigate the effect of annealing on ternary Ni-Fe-Zr-based metallic glasses as a function of temperature and time. We will present results that will show metallic glassy alloys annealed at 973 K for one hour behave like insulators; however, this insulating behavior can become metallic if annealing period at 973 K is longer than four hours.

CP929, *Nanotechnology and Its Applications, First Sharjah International Conference*
edited by Y. I. Salamin, N. M. Hamdan, H. Al-Awadhi, N. M. Jisrawi, and N. Tabet
© 2007 American Institute of Physics 978-0-7354-0439-7/07/$23.00

EXPERIMENTAL

The Zr-Ni-Fe alloy ingots were prepared by arc-melting appropriate amounts of high purity (99.99%) constituent elements in a water-cooled copper hearth with Zirconium getter in an argon atmosphere. The resultant ingots (1.5 grams) were melt-spun under 40 Kpa Ar atmosphere on the surface of a solid 4-inch copper wheel rotating with tangential speed of 35 m/s. The resultant ribbons were typically 1 m long, 1-2 mm wide and 20 μm thick. The amorphous nature of the obtained ribbons was studied by Cu Kα X-ray diffraction. The ribbons were deemed to be amorphous based on the absence of diffraction peaks. A Jeol JSM-5600 scanning electron microscope equipped with Oxford EDX was used to determine the morphology, elemental composition and homogeneity of the produced ribbon. Differential scanning calorimetry (DSC) analyses were carried out in Perkin-Elmer DSC7 under Helium atmosphere at a heating rate of 20 K/min. 5 to 7 cm long ribbons of Ni-Fe-Zr based metallic glass were sealed inside quartz ampoule under vacuum of about 10^{-5} torr. This step was taken to avoid any oxidation of the metallic glass during the heating treatment. The sealed ampoule was then placed inside a Thermolyne 1300 furnace where the desired constant annealing temperature can be set. Different samples of the Ni-Fe-Zr metallic glasses sealed inside quartz ampoules were annealed at 973 K for different periods of time. After the annealing process was finished, the quartz ampoule was allowed to cool down, then it was broken and the annealed sample was collected for further studies. A 4-point dc technique was employed in the I-V and resistivity measurements.

RESULTS AND DISCUSSION

Here we present experimental results and discussions on the effect of the annealing on the temperature dependence of the electrical resistivity in three different metallic glasses with the following compositions: $A=Ni_{0.29}Fe_{0.09}Zr_{0.62}$, $B=Ni_{0.20}Fe_{0.18}Zr_{0.62}$ and $C=Ni_{0.11}Fe_{0.27}Zr_{0.62}$. Figure 1 shows a plot of the relative electrical resistivity ($\rho(T)/\rho(280\ K)$) as a function of temperature for the as quenched metallic glasses A, B, and C. The temperature dependence of the electrical resistivity for the as quenched metallic glasses was also investigated in the presence of constant applied magnetic fields up to 3 kG. The magnetic fields were applied perpendicular and paralle to the direction of the measuring current which was applied along the length of the metallic glassy ribbons; however, there was no noticeable change. The temperature dependence of the electrical resistivity of the as quenched alloys can be well described by the following empirical equation proposed by Mizutani [3]: $\rho(T) / \rho(280\ K) = A + B\ exp(-T/\Delta)$, where $\rho(280\ K)$ is the resistivity at 300 K and A, B and Δ are fitting parameters. This equation is considered for metallic glasses which contain d electrons at the Fermi energy level, E_F. The parameter A is approximately unity and the parameter B ranges from 0.03 to 0.06. The fitted parameter Δ = 220 K, 232 K, 235 K for the alloys A, B and C respectively.

The crystallization temperatures (T_x) obtained from DSC analysis for the three different metallic glasses are 647 K, 660 K, and 655 K for A, B and C respectively. Figure 2 is a plot of the relative electrical resistivity (R(T)/R(280 K)) versus temperature for different samples of the metallic glassy alloys A, B and C annealed at 973 K well beyond their crystallization temperatures. The annealed samples were investigated using EDX analyses, all the samples were found to contain Ni, Fe, and Zr only and their compositions were very close to the compositions of the as quenched metallic glasses.

To investigate the nature of the insulating behavior at low temperatures further, we carried out I-V measurements on the one hour annealed samples. Figure 3 shows the results of I-V measurements performed on a one hour annealed $Ni_{0.11}Fe_{0.27}Zr_{0.62}$ metallic glass. The 4-point I-V measurements were performed at different constant temperatures along the length of the ribbon. The I-V curves display nonlinear behavior at low temperatures from (25 K to ~ 100 K), and the behavior becomes linear for temperatures above 100 K. The I-V measurements were also performed in the presence of applied different magnetic fields. The magnetic fields were applied parallel and perpendicular to the supplied currents along the sample; however, no noticeable effect on the I-V curves was seen. We propose that annealing at 973 K for one hour breaks the entire sample into a large number of nano-crystallites of different Ni, Ni-Zr, Fe, Fe-Zr and Ni-Fe-Zr phases. During the electrical measurements, the applied electric current has to tunnel through the barriers formed by the contact area between the surfaces of these nano-crystallites. As the temperature is lowered further below room temperature, the contact area between the nano-crystallites shrinks and the applied electric current finds it much harder to cross barriers and hence the observed non-linear I-V curves.

124

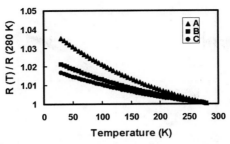

FIGURE 1. The relative electrical resistivity as a function of temperature for the as quenched metallic glasses A, B and C.

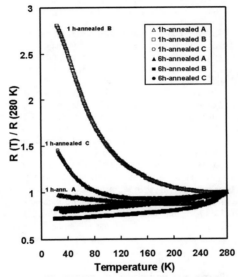

FIGURE 2. The temperature dependence of the relative electric resistivity for three different metallic glasses annealed at 973 K for periods of one and six hours.

Figure 2 clearly shows the effect of annealing, samples annealed for one hour show clear insulating behavior like while samples annealed for six hours show metallic behavior like. The insulating behavior like is very pronounced in the case of the B and C alloys and more so for B than C.

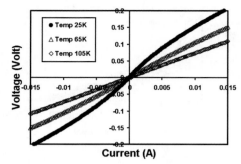

FIGURE 3. Current-Voltage (I-V) curves for a sample of $Ni_{0.11}Fe_{0.27}Zr_{0.62}$ metallic glass annealed at 973 K for one hour, the I-V curves are measured at different temperatures.

The crystallization kinetics in Zr-based metallic glasses is temperature and time dependent in a very complicated way [11-13]. Unfortunately, we do not have TEM facilities at our disposal to track the evolution of the nano-crystallization of our samples and this would be a very interesting research problem. However, the recent work of Liu *et al* on the crystallization of $Fe_{33}Zr_{67}$ and $Fe_{10}Zr_{90}$ metallic glasses [13] is very relevant to the present study. They have found that the crystallization of $Fe_{33}Zr_{67}$ and $Fe_{10}Zr_{90}$ metallic glasses over the temperature range 733 – 1223 K proceeds by forming fine nano-crystallites of different Fe and FeZr phases, and the size of these fine grains grow as a function of annealing temperature and time. The crystallization of $FeZr_2$ metallic glass proceeds by the formation of fcc $FeZr_2$ to fcc $FeZr_2$ + bct $FeZr_2$ and finally stable bct $FeZr_2$ [13], while the crystallization of $NiZr_2$ metallic glass forms a stable bct $NiZr_2$ [12]. We have carried out X-ray diffraction studies to determine the phases present in our crystallized metallic glasses. Figure 4 shows the XRD results of the as quenched and 973 K annealed $Ni_{0.11}Fe_{0.27}Zr_{0.62}$ metallic glass for periods of one and six hours. An amorphous state was obtained for the as quenched $Ni_{0.11}Fe_{0.27}Zr_{0.62}$ metallic glass while diffraction peaks were observed when the present metallic glass was crystallized at 973 K. These diffraction peaks are identified with $NiZr_2$ and $FeZr_2$ crystalline phases. Squares represent bct NiZr2, circles represent $FeZr_2$, and triangles represent fcc $FeZr_2$. The broadening of the diffraction peaks is an indication of smallness of $NiZr_2$ and $FeZr_2$ crystals. The XRD pattern for the one hour annealed and the six hour annealed samples are almost similar except for the six annealed hour sample, the width of the diffraction peaks are about 10-15% narrower in addition to the disappearance of the broad peak seen around $2\theta = 24.5°$ for the one hour annealed sample. In accordance with the Scherrer formula, an average grain size of about 34 nm was estimated from this peak. This peak is identified with fcc $FeZr_2$.

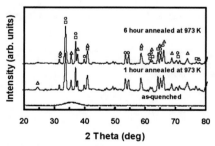

FIGURE 4. XRD patterns of as quenched and crystallized $Ni_{0.11}Fe_{0.27}Zr_{0.62}$ metallic glass at 973 K for periods of one and six hours. Squares (bct $NiZr_2$), circles (bct $FeZr_2$), and triangles (fcc $FeZr_2$).

126

The fcc $FeZr_2$ phase is a metastable phase and it is also known as the big cube structure with 96 atoms [13-14]. Dikeakos *et al* have examined the temperature dependence of the resistivity of the big cube phase over the temperature range 1.6 – 300 K [14]. Their examination revealed a large negative temperature coefficient ($\alpha = (1/\rho) / (d\rho/dT)$). To explain the observed results in this study we propose the following: annealing at 973 K for one hour breaks the entire sample into a large number of fcc $FeZr_2$, bct $FeZr_2$, and bct $NiZr_2$ nano-crystallites. The applied electric current that passes through these nano-crystallites has to tunnel through the barriers formed at their surfaces. As the temperature is lowered, the contact area between the nano-crystallites shrinks and the applied electric current finds it much harder to cross barriers as it is evident from the results of figure 3. The observed insulting behavior like is due to the presence of fcc $FeZr_2$ and the tunneling effects combined. As the period of the annealing is increased the fcc $FeZr_2$ phase transforms into more stable bct $FeZr_2$ crystallites in addition to the coalescence the nano-crystallites forming bigger crystallites. The applied electric current now finds easier flow channels and hence the metallic behavior like observed for the six hour annealed samples. We would like to think that in the case of the one hour 973 K annealed $Ni_{0.29}Fe_{0.09}Zr_{0.62}$ metallic glass we have less fcc $FeZr_2$ than the other two metallic glasses, but in the case of the one hour 973 K annealed $Ni_{0.20}Fe_{0.18}Zr_{0.62}$ metallic glass there is a big competition between the formation of fcc $FeZr_2$, bct $FeZr_2$, and bct $NiZr_2$ nano-crystallites and the tunneling effects become more dominant than the case of the one hour 973 K annealed $Ni_{0.11}Fe_{0.27}Zr_{0.62}$ metallic glass.

CONCLUSION

Three different metallic glasses with the following compositions: $Ni_{0.29}Fe_{0.09}Zr_{0.62}$, $Ni_{0.20}Fe_{0.18}Zr_{0.62}$ and $Ni_{0.11}Fe_{0.27}Zr_{0.62}$ were prepared and characterized. The temperature dependence of the electrical resistivity of the as quenched metallic glasses can be well described by Mizutani's empirical equation for which the Fermi level is in the d band. 5 to 7 cm long ribbons of the three different metallic glasses were sealed inside quartz ampoules under vacuum. The sealed metallic glasses were annealed at 973 K for periods of one and six hours. The annealing process seems to produce different structural environments which manifest themselves by the observation of different electrical behaviors. Insulting electrical behavior like and nonlinear *I-V* characteristics were observed for the one hour 973 K annealed samples while Metallic behavior like is seen for the six hour 973 K annealed samples. We have attributed the observed insulting electrical behavior like and nonlinear *I-V* characteristics to the presence of the metastable fcc $FeZr_2$ nano-crystallites and the tunneling effects between the large number of fcc $FeZr_2$, bct $FeZr_2$, and bct $NiZr_2$ nano-crystallites within the nano-crystallized Ni-Fe-Zr metallic glasses.

REFERENCES

1. P. Duwez, R. H. Willens and W. Klement, J. Appl. Phys. **31**, 1136-1137 (1960).
2. U. Mizutani, Prog. Mater. Sci. **28**, 97-228 (1983).
3. U. Mizutani in: S. Steeb, H. Warlimont (Eds). Proceedings of the Fifth International Conference on Rapidly Quenched Metals. (Elsevier, Amsterdam., 1984) pp. 977.
4. U. Mizutani, Mater. Sci. Eng. **99**, 165-178 (1988).
5. U. Mizutani, M. Tanaka and H. Sato, J. Phys. F: Met. Phys. **17**, 131-141 (1987).
6. Z. Altounian, S. V. Dantu and M. Dikeakos, Phys. Rev. B **49**, 8621-8626 (1994).
7. F. Hamed, F. S. Razavi, S. K. Bose and T. Startseva, Phys. Rev. B **52**, 9674-9678 (1995).
8. K. Hono, Prog. Mater. Sci. **47**, 621-729 (2002).
9. K. Tadeusz, J. Non-Crys. Solids **287**, 145-161 (2001).
10. M. T. Clavaguera, N. Clavaguera, D. Crespo and T. Pradell, Prog. Mater. Sci. **47**, 559-619 (2002).
11. J. Saida, C. Li, M. Matsushita and A. Inoue, Appl. Phys. Lett. **76**, 3037-3039 (2000).
12. Z. Altounian, E. Batalla, J. O. Ström-Olsen and J. L. Walter, J. Appl. Phys. **61**, 149-155 (1987).
13. X. D. Lui, X. B. Liu, and Z. Altounian, J. Non-Crys. Solids **351**, 604-611 (2005).
14. M. Dikeakos, Z. Altounian, and M. Fradkin, Phys. Rev. B **70**, 024209-1-024209-6 (2004).

Synthesis and Thermoluminescence of ZnS:Mn²⁺ Nanoparticles

M. Zahedifar[a], N. Taghavinia[b] and M. Aminpour[a]

[a]*Physics Department, University of Kashan, Kashan, Iran*
[b]*Physics Department, Sharif University of Technology, Tehran, Iran*
zhdfr@kashanu.ac.ir

Abstract. The controlled chemical method has been used for synthesis of Mn doped ZnS nanoparticles. Optical absorption studies showed that increasing of surfactant density, from 0.0001 to 0.5 mol/lit., causes the size of nanoparticles to decrease from 4.8 nm to about 3 nm and the band gap width to increase from 4.15 to 4.50 eV. Also increasing the temperature during the synthesis process caused the nanoparticle size to be increased. As a new result we observed a thermoluminescence (TL) glow peak at about 475 K, with its intensity depending on concentration of the Mn dopant. Activation energy of this glow peak was obtained to be about 0.6eV. A discussion of the obtained results is also presented.

Keywords: ZnS:Mn²⁺, synthesis, thermoluminescence.
PACS: 78.67.-n; 78.60.-b; 78.60.Kn

INTRODUCTION

In nanometer sized particles, with decreasing of particle size, high surface area to volume ratio is obtained which leads to an increase in surface specific active sites and enhanced absorption efficiency[1]. The surface states are very important to the physical properties, especially the optical properties of the nanoparticles. Luminescence efficiency of ZnS: Mn²⁺ nanoparticles has extensively been investigated. Increasing of quantum efficiency with decreasing of particle size has been reported by Bhargava[2]. Also it has been found that the quantum efficiency of nanocrystalline ZnS:Mn²⁺ depends on Mn²⁺ concentration[3]. Besides, the synthesis conditions play a deterministic role in the surface defects and enhancing the efficiency. Thus, the study of surface states or trapping states is also of great importance. Many researchers have devoted their attention to synthesis and optical properties of ZnS nanoparticles[4-6] but the system of recent interest has been Mn²⁺ doped ZnS nanoparticles, due to its interesting electro-optical[7] and magnetic[8] properties. Thermoluminescence (TL), besides its extensive applications in dosimetry and dating, is an efficient tool for detection of traps and determination of physical properties of trapping states both in bulk and nanoscaled materials[9]. In this paper we have presented the process of synthesis of ZnS:Mn²⁺ nanocrystalline and for the first time reported its thermoluminescence and trapping parameters.

RESULTS AND DISCUSSIONS

Doped ZnS:Mn²⁺ nanoparticles were synthesized using a chemical precipitation method. The precipitation was performed starting with a homogeneous solutions of zinc nitrate $Zn(NO_3)_2.6H_2O$ with concentration of 20 mMol/lit and manganese nitrate ($Mn(NO_3)_2.4H_2O$) with concentration of 1mMol/lit and thioglycerol (TG) as surface active agent with different concentrations in 50cc DI water as solvent and sodium sulfide ($Na_2S.8H_2O$) with concentration of 10mMol/lit in 25cc DI water . Then with holding the system at a constant temperature the effect of TG concentration on

CP929, *Nanotechnology and Its Applications, First Sharjah International Conference*
edited by Y. I. Salamin, N. M. Hamdan, H. Al-Awadhi, N. M. Jisrawi, and N. Tabet
© 2007 American Institute of Physics 978-0-7354-0439-7/07/$23.00

bsorption, band gap and optical size of ZnS:Mn^{2+} nanoparticles were studied. The TG molecules with covering the ZnS:Mn^{2+} clusters controls the size of nanoparticles. At the end, with adding of methanol or acetone to the solution, nanoparticles precipitated and the white powder were removed by centrifuging and washing the solution. Increasing the temperature causes more decomposition of sodium sulfide (Na$_2$S) and thus increased rate of ZnS:Mn^{2+} production. This effect results in a lower band gap showing the increased size of nanoparticles. As the first method, UV-Vis system model Jasco V-530, was used for optical absorption spectroscopy. By investigation of transmitted light[10] the band gap was evaluated and then the particle size was determined using effective mass approximation[11]. As can be observed in figure 1 with increasing the surfactant concentration from 0.0001 Mol/lit to 0.5 Mol/lit, the absorption shifts to smaller wavelengths. Also the effect of temperature and time duration of synthesis for different surfactant concentrations was studied. It was found that increasing of both the temperature and the time duration cause an increase in particle size. The X-ray diffraction spectroscopy was carried out using Bruker X-ray diffractometer. The size of nanoparticles was determined by Scherer formula $d = 0.94\lambda /(\beta_{hkl} \cos\theta)$ which gives the particle size using β_{hkl} (the FWHM for diffraction from crystalline surface hkl). Figure 2 shows the XRD diffraction patterns of three synthesized samples in different temperatures and time durations of 1 hour at 95°C(a), 30 minute at 95°C (b) and at room temperature (c).

The concentration of surfactant in three samples was 0.0001 mol/lit. As can be observed, the heating and the time duration have important effects on crystalline formation. This property was observed for different concentrations of surface active agent. The particle sizes using Scherer formula were determined to be 5.1, 4.5 and 3.8 nm from patterns a, b, and c, respectively, which can be compared to the corresponding optical sizes of 5.62, 5 and 4.13 nm. The Scherer formula gives more accurate results compared to those of the effective mass approximation.

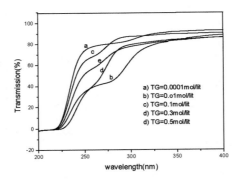

FIGURE 1. The transmission spectra from ZnS:Mn^{2+} for different concentrations of surfactant and 1 hour heating at 95°C.

Thermogravimetry analysis (TGA), gives us information about the variation of mass of sample as a function of temperature. This analysis is important in studying the thermal stability, especially in thermoluminescence (TL) analysis which needs to raise the temperature for recording the glow curve. In TGA analysis 10.84 mg of synthesized ZnS:Mn^{2+} nanoparticles was heated from room temperature up to 500°C with the rate of 10°C/sec.

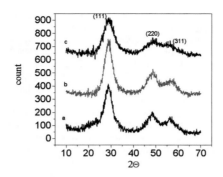

FIGURE 2. XRD diffraction pattern for three different heating conditions of (a) 1 hour at 95°C (b) 30 min at 95°C, and (c) without heating. TG concentration is 0.0001 mol/lit.

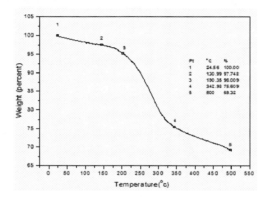

FIGURE 3. TGA of synthesized nanocrystalline. Points 1through 5 refer to different weights of the sample at different temperatures.

Figure 3 shows that for temperatures up to 250°C the reduction of mass and thus variation of crystalline structure can be ignored. For TL analysis, the synthesized sample was exposed to β particles from a ^{90}Sr-^{90}Y source from which the sample received a dose of 350 mrad. A TL reader model 4500 from Harshaw was used for recording the TL glow curves. A computer program for glow curve deconvolution and obtaining the kinetics parameters (which has been developed in our laboratory based on the Levenberg-Marquart algorithm) was used for curve fitting and kinetic analysis. For testing the goodness of fit, the figure of merit (FOM) has been used.[12]. The general-order model for TL was used for kinetic analysis of glow curves. Two samples with the same conditions of synthesis (TG concentration of 0.4mol/lit and 1 hour heating at 95°C), but different concentrations of 10 mMol/lit and 1mMol/lit of $Mn(NO_3)_2.4H_2O$ were used for TL analysis. The optical sizes of two samples were determined to be 2.8 nm and 4.13 nm respectively.

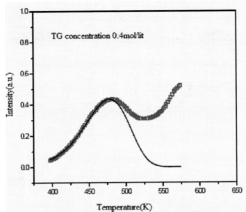

FIGURE 4. TL glow curve (open rectangles) and deconvoluted glow peak (solid line) due to Mn dopant.

Figure 4 shows the TL glow curve and the deconvoluted glow peak of the sample with higher dopant ratio. This glow peak at about 475 K has resulted solely due to the Mn dopant. The FOM was obtained to be 0.82% and 2.2% for higher and lower Mn dopant, respectively, which shows a good fit.

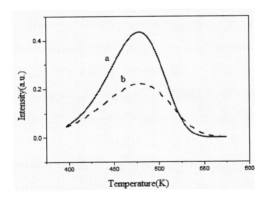

FIGURE 5. Deconvoluted glow peaks of ZnS;Mn^{2+} nanoparticles due to Mn dopant.

In figure 5 two glow peaks corresponding to different dopant concentrations are shown. As can be seen, with increasing dopant ratio, the TL intensity increases which can be attributed to increasing the trapping states with increasing Mn concentration. Also our analysis showed a first–order behavior for TL glow peaks. The activation energies of samples with high and low dopant concentrations were obtained to be 0.61eV and 0.58 eV, respectively, which lie within the systematic errors.

REFERENCES

1. A. J. Hoffman, G.Millis, H. Yee and M. R. Hoffman, J. Phys. Chem. 96, 5546 (1992).
2. R. N. Bhargava and D. Gallagher, Phys. Rev. Lett.72, 416 (1994).

3. K. Sooklal,B. S. Cullum, S. M. Angel and C. J. Murphy, J. Phys. Chem. **100**, 455 (1996).
4. A. S. Ethiraj, N. Hebalkar and S. K. Kulkarni, J. Chem. Phys. **118**, 8945 (2003).
5. W. Chen, Z. Wang, Z. Lin and L. Lin, J. Appl. Phys. **82**, 3111 (1997).
6. V. Turko Liver, M. Ross, G. D. Arrigo, D. Marin and G. Mircocci, Applied Physics A **69**, 369 (1999).
7. R. N. Bhargava, J. Lumin. **70**, 85 (1996).
8. T. A. Kennedy, E. R. Glasser, P. B. Klein and R. N. Bhargava, Phys. Rev. B **52**, 356 (1995).
9. W. Cheng, Z. Wang, Z. Lin and L. Lin, Appl. Phys. Lett. **70**, 11 (1997).
10. R. Viswanatha, S. Sapra, S. S. Gupta, B. Satpati, P. V. Satyam, B. N. Dev and D. D. Sarma, J. Phys. Chem. B **108**, 6303 (2004).
11. B. A. Smith, J. Z. Zhang, A. Joly and J. Lin, Phys. Rev. B **62**, 2021 (2004).
12. H. G. Balian and N. W. Eddy, Nucl. Instr. and Meth. **145**, 389 (1977).

Preparation of Nano-Scale Magnetite Fe₃O₄ and its Effects on the Bulk Bi-2223 Superconductors

N. Ghazanfari [a], A. Kılıç [a,e], Ş. Özcan [b], H. Sözeri [c], H. Özkan [a] and A. Gencer [d]

[a]METU, Department of Physics, Ankara Turkey
[b]Hacettepe University, Department of Physics Engineering, Ankara Turkey
[c] TUBITAK-UME, National Metrology Institute, Kocaeli, Turkey
[d]Ankara University, Department of Physics, Ankara Turkey
[e]Niğde University, Department of Physics, Niğde Turkey
hozkan@metu.edu.tr

Abstract. Nano-scale magnetite (Fe_3O_4) powders were prepared from metallic iron and distilled water by wet milling technique. Magnetite powders so obtained were added to $Bi_{1.6}Pb_{0.4}Sr_2Ca_2O_y$ superconductor by 0.00, 0.05, 0.10 and 0.30 wt % (x), with solid state reaction method. The structural and superconducting properties were studied by x-ray diffraction, transmission electron microscopy, magnetization and magnetic susceptibility measurements. Single-phase magnetite samples with average particle size about 25 nm were obtained with the wet milling technique. Nanoscale magnetite additions up to 0.10 wt % enhance the critical current density (J_c) of the superconductor preserving the fraction of the high T_c phase Bi-2223. Magnetite additions above 0.10 wt % decrease the critical parameters (T_c, J_c) and the fraction of the high T_c phase.

Keywords: Magnetite Fe_3O_4 nano-particles, TEM, Bi-2223, Critical current, AC susceptibility.
PACS: 74.25.Sv, 74.25.Ha, 74.72.Hs

INTRODUCTION

Magnetic nanoparticles are important for optoelectronic and magnetic devices. Several methods have been used to produce nano-scale Fe_3O_4 [1-3]. We have prepared rather small magnetite particles with the wet-milling technique. This new method of magnetite preparation is described below.

High temperature superconductors (HTSs) are not suitable for making wires and tapes due to their granular microstructure. Inter-grain week links diminish the critical current density (J_c). HTSs are Type II materials; motions of vortices create resistance and cause energy dissipation. These facts decrease j_c and restrict their applications. Structural modifications to improve the properties have been subject of continued interests. Nanoparticles addition during their synthesis may create defects and pining centers for J_c enhancements. The sizes and densities of defects are among the critical factors to be adjusted for effective vortex pining. SiC nanoparticles in MgB_2 [4] and MgO, ZrO_2 and Al_2O_3 nanoparticles in Bi-based superconductors (BSCCO) increase their J_c [5-7]. Few papers available in the literature about Fe substitutions in BSCCO report decrease of T_c without specifying the sizes of Fe particles [8, 9]. Recently, it was shown that magnetic nanoparticles as Fe_2O_3 embedded into the bulk Type II superconductors act as efficient pinning centers [10, 11]. There is no report in the literature about iron oxides additions on properties of BSCCO. We have added nanoparticles of Fe_3O_4 to $Bi_{1.6}Pb_{0.4}Sr_2Ca_2Cu_3O_{10}$ superconductor during the synthesis. This paper reports the preparation of magnetite Fe_3O_4 nanoparticles and their effects on the formation and properties of BSCCO.

CP929, Nanotechnology and Its Applications, First Sharjah International Conference
edited by Y. I. Salamin, N. M. Hamdan, H. Al-Awadhi, N. M. Jisrawi, and N. Tabet
© 2007 American Institute of Physics 978-0-7354-0439-7/07/$23.00

EXPERIMENTAL

Metallic Fe of average particle size less than 30 µm were mixed with distilled water with a molar ratio of 3(Fe):4(H$_2$O). The mixture was milled with steel balls in a stainless steel vial for 22 h at 350 rpm. After first milling, the precursor powder in a porcelain crucible was heated at 700 °C for 1 h and furnace cooled. The calcined powder was milled again for 6 h at a milling rate of 350 rpm to reduce the particle size further. The samples were characterized by XRD and SEM.

Appropriate amounts of pure oxides Bi$_2$O$_3$, PbO, SrO, CaCO$_3$, and CuO were mixed and calcined at 800 °C for 24 hours to lead the Bi$_{1.6}$Pb$_{0.4}$Sr$_2$Ca$_2$Cu$_3$O$_{10}$. Then, the precursor was mixed with Fe$_3$O$_4$ in weight percents (wt %); x = 0.00, 0.05, 0.10, and 0.30. The samples were pressed into pellets, calcined twice at 840 °C and 850 °C for 96 h and 120 h, respectively.

X-ray diffraction patterns (XRD) with Rigaku Miniflex Diffractometer (CuK$_\alpha$ radiation) were obtained for phase identifications. Transport measurements were carried out with four-probe method. The samples were cooled in a liquid nitrogen cryostat with N$_2$ gas as heat exchanger. Inductive critical current densities were calculated by the Bean Model from the magnetization measurements. Susceptibility measurements were performed to study inter-and intra-grain features.

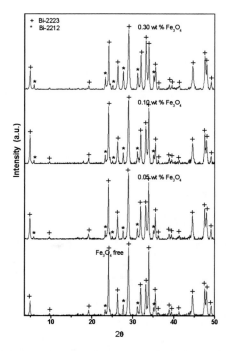

FIGURE 1. X-ray diffraction patterns of the BSCCO samples. The peaks of the high and low-T$_c$ phases (Bi-2223 and Bi-2212) are indexed as + and *, respectively.

134

RESULTS AND DISSCUSION

XRD patterns show that single magnetite Fe_3O_4 phase was formed with the wet milling technique. The diffraction data agree well with the ICDD standards. Average particle sizes of Fe_3O_4 were estimated to be about 25 nm with the electron micrographs and the Scherrer formula.

XRD patterns of the pure and Fe_3O_4 added BSCCO samples are shown in Figure 1. The quantitative analyses carried out with the Miniflex Diffractometer Software are listed in Table 1. The samples are mixtures of Bi-2212 and Bi-2223 phases, the fraction of the later being more than 94 % for x up to 0.10.

TABLE 1. Volume fractions of the Bi-2212 and Bi-2223 phases.

	x = 0.00	x = 0.05	x = 0.10	x = 0.30
Bi-2212	4	5	6	10
Bi-2223	96	95	94	90

Resistance versus temperature (R-T) plots of the BSCCO samples were shown in Figure 2. These figures show sharp transition for x up to 0.10. The transition width increase from 4-5 K to about 10 K for x = 0.30. The critical temperatures (T_c) of the samples determined from the R – T plots are shown in figure 3. T_c drops from 106 K to 99 K with increase of x up to 0.30.

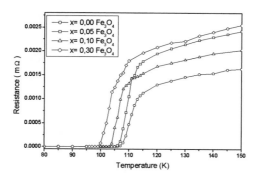

FIGURE 2. Electrical resistance vs. temperature for the pure and Fe_3O_4 added BSCCO samples.

The critical current densities (J_c) at 50 and 77 K calculated from magnetization data with the Bean model are shown in figure 4. J_c values increase up to 40 % with Fe_3O_4 nanoparticles additions up to 0.10 wt %. Significant enhancements of J_c suggest improvements of inter-grain connections with Fe_3O_4 additions.

The real and imaginary parts of ac susceptibility (χ', χ'') vs. temperature for ac fields of 20 A/m with f= 111 Hz are shown in figure 5. χ'-T plots show two sharp drops below T_c for x up to 0.10. The drops occurred at lower temperatures for x = 0.30. These observations indicate that intra-grain transitions were not effected much with Fe_3O_4 nanoparticles additions up to 0.10 wt %.

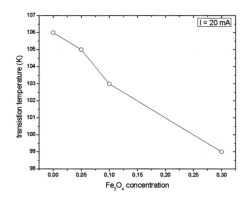

FIGURE 3. Critical temperature vs. Fe_3O_4 concentration for the pure and magnetite added samples.

The χ''-T plots shift to lower temperatures with Fe_3O_4 nanoparticles additions, the shift being least for x = 0.10 and most for x = 0.30. These observations indicate that Fe_3O_4 nanoparticles additions do not significantly affect the ac losses for x up to 0.10. The larger shift for the samples for x = 0.30 suggest that dissipations and field penetrations increase with increase of Fe_3O_4 concentration.

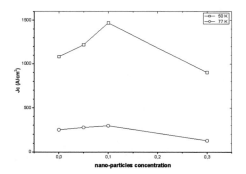

FIGURE 4. Inductive critical current density vs. Fe_3O_4 concentration at 50 and 77 K.

CONCLUSION

Nano-scale magnetite Fe_3O_4 powders were prepared from metallic iron and distilled water by the wet milling technique and added to BSCCO. Fraction of the Bi-2223 phase was preserved with Fe_3O_4 nanoparticles additions up to x = 0.10 wt %. T_c values decrease up to x = 0.30 and the J_c values increase up to x = 0.10 and then they decrease with further increase of x. Intra-grain transitions and ac losses change little with Fe_3O_4 additions up to x = 0.10. J_c enhancements suggest improvements of inter-grain connections with Fe_3O_4 nanoparticles additions up to x = 0.10.

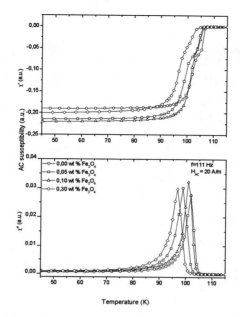

FIGURE 5. AC susceptibility vs. T for the pure and Fe_3O_4 added BSCCO samples

ACKNOWLEDGMENTS

This work was supported by TÜBİTAK project No: 106T039. One of the authors (A K) thanks to TÜBİTAK, for 2218 Programme support.

REFERENCES

1. G. F. Goya, Solid State Commun. **130**, 783 (2004).
2. G. F. Goya, T. S. Berquo, F. C. Fonseca and M. P. Morales, J. App. Phys. **94**, 3520 (2003).
3. J. Wang, K. Zhang, Z. Peng and Q. Chen, J. Cryst. Growth **266**, 500 (2004).
4. SX. Dou, S. Soltanian, J. Hovat, X. L. Wang, SH. Zhou, M. Ionescu, H. K. Liu, P. Minroe, M. Tomsic, Apl. Phys. Lett. **81**, 3419-3421 (2002).
5. W. D. Huang, W. H. Song, Z, Cui, B. Zhao, M. H. Pu, X. C. Wu, Y. P. Sun, and J. J. Du, Phy. Stat. Sol. (a) **179**, 189 (2000).
6. Z. Y. Jia, H. Tang, Z. Q. Yang, Y. T. Xing, Y.Z. Wang, and G.W. Qiao, Physica C **337**, 130 (2000).
7. M. Annabi, A. M'chirgui, F. Ben Azzouz, M. Zouaoui, M. Ben Salem, Physica C **405**, 25 (2004).
8. A. V. Pop, GH Ilonca, D. Ciurchea, M. Ye, I. I. Geru, V. G. Kantser, V, Vop, M. Todic and Deltour, Journal of Alloys and Compounds **241**, 116 (1996).
9. P. Sumana Praabu, M. S. Ramachandra Raao and G. V. Subba Rao, Physica C **211**, 279 (1993).
10. R. Prozorov, T. Prozorov, A. Snezhko IEEE Transitions on Applied Superconductivity **15**, 3277 (2005).
11. A. Snezhko, T. Prozorov, R. Prozorov, Phys. Rev. B **71**, 024527 (2005).

137

Effect of Nb Addition on the Critical Parameters of BSCCO Superconductor

H. Sozeri[1], N. Ghazanfari[2], H. Özkan[2], A. Kilic[2,3] and N. Hasanli[2]

[1] *TUBITAK-UME, National Metrology Institute, 41470 Gebze Kocaeli, TURKEY*
[2] *Middle East Technical University, Physics Department, 06531 Ankara, TURKEY*
[3] *Nigde University, Physics Department, 51200 Nigde, TURKEY*
hozkan@metu.edu.tr

Abstract. Abstract. Pure and Nb_2O_5 added $Bi_{1.6}Pb_{0.4}Nb_xSr_2Ca_2Cu_3O_8$ superconductors ($0 \leq x \leq 0.30$) were synthesized by the ammonium nitrate melt route. Fraction of the high T_c 2223 phase increases with addition of small amounts of Nb. AC susceptibility and transport measurements showed the optimum Nb concentration as $x = 0.10$. Nb additions above $x = 0.20$ decrease significantly the fraction of the high-T_c phase and degrade the superconducting properties.

Keywords: Ceramic BSCCO; Ammonium nitrate melt; Transport properties
PACS: 74.25 Ha; 74.25 Sv; 74.25 Fy.

INTRODUCTION

The three superconducting phases in Bi-Sr-Ca-Cu-O system (BSCCO) can be represented by an ideal formula $Bi_2Sr_2Ca_{n-1}Cu_nO_{4+2n}$, where n = 1, 2 or 3 indicate the number of CuO_2 planes. Each phase has different critical temperature: n = 1 (2201) T_c = 20 K, n = 2 (2212) T_c = 80 K and n = 3 (2223) T_c = 110 K. The 2212 and 2223 phases are important for technological applications. But, it is rather difficult to obtain them in isolated form as the 2212 phase grows prior to the 2223 phase during the synthesis. Since the discovery of BSCCO superconductor, many substitutions have been tried to improve formation and stability of the high-T_c 2223 phase [1, 2]. Among these, partial substitution of Bi with Pb has been found to be most effective one. The nominal concentration of Pb is determined as 0.4. There are contradicting reports in the literature about the effects of Nb doping on the formation and properties of BSCCO [2-5]. Nasu *et al.* [4] observed that both T_c and volume fraction of the 2223 phase increases as Nb concentration raises up to x = 0.2 in $Bi_{1.6}Pb_{0.4}Nb_xSr_2Ca_2Cu_3O_8$. They related the increase of T_c to the change of melting points caused by Nb additions.

In this work, Nb_2O_5 was added to $Bi_{1.6}Pb_{0.4}Nb_xSr_2Ca_2Cu_3O_8$ superconductor with concentrations between x = 0 and x = 0.30. We aimed to study the evaluation of the 2223 phase and the superconducting properties with Nb addition.

EXPERIMENTAL

We have prepared $Bi_{1.6}Pb_{0.4}Nb_xSr_2Ca_2Cu_3O_8$ samples with Nb content between x = 0 and 0.30 with the ammonium nitrate melt (ANM) route. Detailed description of the ANM method can be found in Ref. [6, 7]. XRD patterns obtained by the Rigaku Miniflex powder diffractometer with CuK_α radiation were used to determine the contents of the phases. Transport measurements with four probe method were carried out to obtain the transition temperatures. Inductive critical current densities were calculated using the Bean model [8] from the magnetization measurements that were

CP929, *Nanotechnology and Its Applications, First Sharjah International Conference*
edited by Y. I. Salamin, N. M. Hamdan, H. Al-Awadhi, N. M. Jisrawi, and N. Tabet
© 2007 American Institute of Physics 978-0-7354-0439-7/07/$23.00

carried out using SQUID magnetometer MPMS-XL. AC susceptibility measurements were performed using a Lake Shore 7130 susceptometer and a closed-cycle refrigerator at low temperatures down to 15 K.

RESULTS AND DISCUSSIONS

XRD patterns of the Nb free and Nb_2O_5 added samples synthesized with ANM method demonstrated the phase formations and intensity changes with Nb additions. All the samples are superconducting with traces of Ca_2PbO_4. The fractional amounts of the 2223 and 2212 phases were calculated using the quantitative analysis software of the Rigaku diffractometer. The results are listed in Table 1. Nb free samples have both 2223 and 2212 phases, fraction of latter being higher. As the Nb concentration increases above $x = 0.05$ the fraction of the 2212 phase decreases and that of the 2223 phase increases. But, for $x = 0.30$ the fraction of the impurity phases become significant.

TABLE 1. High- and low-T_c phase fractions of the samples together with T_c offset values. Besides 2223 and 2212, Ca_2PbO_4 was also detected and included into other phase's column.

Nb concentration	2223 (%)	2212 (%)	Oth. (%)	T_c (K)
$x = 0.00$	46	51	3	108
$x = 0.05$	50	50	0	110
$x = 0.10$	58	40	2	108
$x = 0.20$	55	40	5	101
$x = 0.30$	45	10	45	< 77

The DC electrical resistances versus temperature $(R–T)$ graphs of the samples are shown in Fig. 1. T_c offset determined from $R–T$ graphs are listed in Table 1 and plotted versus Nb concentration in Fig. 2. T_c of the samples remains above 100 K for x up to 0.20 and then it rapidly decreases for $x = 0.30$.

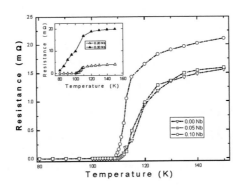

FIGURE 1. Resistance vs. temperature curves for the samples with different Nb concentration.

Inductive critical current densities (J_c) versus x (Nb content) at $T = 50$ and 77 K are presented in Fig. 3. The features of J_c versus x graphs are as follows: J_c values increase for x up to 0.10 and then sharply decrease with increase of Nb content above $x = 0.20$. The increase of J_c indicates that the weak links and vortex pinning improve with Nb addition up to $x = 0.10$. The ANM process leads to smaller grains, more grain boundaries and weak links, as we have noticed in the previous report [7]. Increased Nb contents for $x > 0.20$ appear to degrade coupling between grains and decrease the critical parameters.

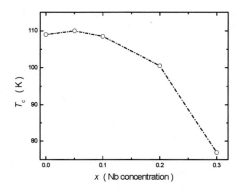

FIGURE 2. T_c values as a function of Nb concentration.

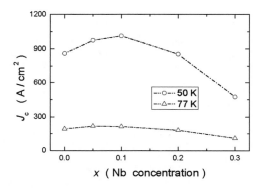

FIGURE 3. Critical current density is shown as a function of Nb concentration.

Figure 4 shows the real (χ') and imaginary (χ'') parts of AC susceptibility versus temperature for Nb free and Nb added samples for fields of 20 A/m with f = 111 Hz. All the measurements reported here are normalized to the value of χ' at 45 K. The samples studied show two significant drops in $\chi' - T$ graphs as temperature is decreased. These are related to intra-grain (first one near T_c) and inter-grain (second one) transitions. The intra-grain and inter-grain transition temperatures (T_{tr}) of Nb free and Nb added samples are obtained from the derivatives of $\chi' - T$ graphs (see Fig. 5) and listed in Table 2.

TABLE 2. Intra- and inter-grain transition temperatures T_{tr} (K) of the samples.

Nb concentration	Intra-grain	Inter-grain
$x = 0.00$	106.76	100.71
$x = 0.05$	106.14	101.10
$x = 0.10$	106.08	101.30
$x = 0.20$	105.17	93.26
$x = 0.30$	104.86	59.90

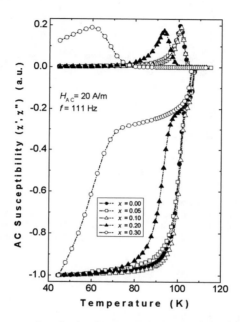

FIGURE 4. Temperature dependencies of the real and imaginary parts of the AC susceptibility.

The compositional variation of the inter-grain transition temperatures, revealed from $\chi' - T$ data, are in good agreement with that of T_c values obtained from $R-T$ data. Intra-grain transition temperature does not change much with Nb addition. The inter-grain transitions are quite sharp for the samples with $x < 0.20$. At $x \geq 0.20$, the inter-grain transition temperatures drastically decrease with increase of Nb content, as also was observed for T_c from $R-T$ data. The maximum in the imaginary part of susceptibility versus temperature ($\chi'' - T$) graph is related to effective flux penetration and grain coupling. Such peaks, near T_c are quite sharp for the Nb containing samples ($x \leq 0.10$) indicating that low levels of Nb additions do not degrade couplings between the grains in the samples.

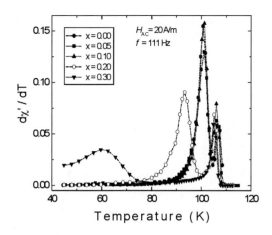

FIGURE 5. Temperature dependencies of first derivative of real part of the AC susceptibility.

CONCLUSIONS

$Bi_{1.6}Pb_{0.4}Nb_xSr_2Ca_2Cu_3O_\delta$ superconductor was sintered by ANM route with Nb concentrations varying from $x = 0$ to 0.30. It was observed that Nb addition slightly enhances the high-T_c 2223 phase formation and improves the critical parameters T_c and J_c. The fraction of the 2223 phase increases up to Nb concentration of $x = 0.10$. Further increase of Nb content decreases both T_c offset and J_c. AC susceptibility and transport measurements showed that optimum Nb concentration is $x = 0.10$.

REFERENCES

1. H. Maeda, Y. Tanaka, M. Fokutomi and T. Asano, *Jpn. J. Appl. Phys.* **27**, L209 (1988).
2. T. Kanai, T. Kamo and S. Matsuda, *Jpn. J. Appl. Phys.* **28**, L551 (1989).
3. Y. Li and B. Yang, *J. Mater. Sci. Lett.* **13**, 594 (1993).
4. H. Nasu, N. Kuriyama and K. Kamiya, *Jpn. J. Appl. Phys.* **29**, L1415 (1990).
5. D. R. Mishra, P. L. Upadhyay and R. G. Sharma, *Physica C* **304**, 293 (1998).
6. G. K. Strukova *et al.*, *Supercond. Sci. Tech.* **6**, 589 (1993).
7. H. Sozeri, H. Ozkan and N. Ghazanfari, *J. All. Comp.* **428**, 1 (2007).
8. C. P. Bean, *Rev. Mod. Phys.* **36**, 31 (1964).

Effects of Al$_2$O$_3$ Nano-Particles on the Irreversible Properties of MgB$_2$ Superconductor

Kh. A. Ziq[1], M. Shahabuddin[2,3], I. Ansari[3], A. F. Salem[1], K. Singh[4] and H. Kishan[4]

[1]*Department of Physics, King Fahd University of Petroleum and Minerals, Dhahran, Saudi Arabia*
[2]*Department of Physics, King Saud University, Riyadh, Saudi Arabia*
[3]*Department of Physics, Jamia Millia Islamia, 110025 New Deli, India*
[4]*National Physical Laboratory, K. S. Krishnan Marg, 110012 New Delhi, India*
kaziq@kfupm.edu.sa

Abstract. Magnetic measurements on MgB$_2$ superconductors doped with various concentration (0 < x < 0.6%) of Al$_2$O$_3$ nano-particles (~20 nm) have been performed in the temperature range 4-40K and in a magnetic field of strength up to 9 Tesla. A significant increase in the irreversibility field (Hirr), critical current density (Jc) and remanent magnetization (M$_R$) have been obtained with increasing the concentration of the Al$_2$O$_3$ nano-particle. At low field we have observed large vortex instabilities (known as vortex-avalanche) associated with all doped samples characterized with high critical current density. Vortex avalanche effect is reduced with increasing temperature vanishing near 20K. The results are discussed in terms of local-vortex instabilities caused by doping of Al$_2$O$_3$ nano-particles.

Keywords: vortex motion, vortex instability.
PACS: 61.46Df; 75.75.ta; 74.25.Qt.

INTRODUCTION

Since the discovery of superconductivity at 39K in MgB$_2$ [1], major efforts have been devoted towards understanding basic superconducting and normal state properties of this binary inter-metallic compound. However, the substitution chemistry in MgB$_2$ has proven not to be so simple. Limited substitution on either the Mg or B sites has been achieved for only a few elements-for example, Al on the Mg site or C on the B site [4].

Chemical substitution has been found to affect various superconducting and normal state properties of MgB$_2$ material [2].

In this paper, we are introducing Al$_2$O$_3$ nano-powder, not as a substitution but as a modifier of the grain boundaries, hence, affecting the coupling between the grains and many other related properties, such as critical current density and instability of the vortex structure.

EXPERIMENTAL

Solid state reaction has been followed in preparing MgB$_2$ samples. High purity (>99.9%) Mg, B powders were mixed in stoichiometric ratio along with 20 nm Al$_2$O$_3$ nanoparticles (up to 6% Al$_2$O$_3$ has been used). The combined materials were encapsulated in soft Fe-tubes, evacuated and sealed in quartz tubes under vacuum. The capsules were introduced to a preheated furnace at 750 C, and kept for an optimal time of 2.5 hours, then quenched to liquid nitrogen temperature [5].

CP929, *Nanotechnology and Its Applications, First Sharjah International Conference*
edited by Y. I. Salamin, N. M. Hamdan, H. Al-Awadhi, N. M. Jisrawi, and N. Tabet
© 2007 American Institute of Physics 978-0-7354-0439-7/07/$23.00

Resistivity measurements were performed using standard four probe technique in a closed cycle refrigerator. Magnetic measurements were carried out on a 9-Tesla PAR-4500 vibrating sample magnetometer, in the temperature range 4-50K. Temperature was monitored using C-glass thermometer with accuracy better than 0.05K. The noise level in the magnetization is ~10^{-5} emu.

RESULTS AND DISCUSSIONS

Electron microscopy micrographs of the prepared samples have been taken at room temperature. The results are presented in figure 1 for the sample with 6% Al_2O_3 weight ratio. The micrograph shows good connectivity of the grains, and relatively large grain size, with lots of porosities (Fig.1A). A close-up micrograph Fig. 1B, shows a uniform dispersion of the Al_2O_3 particles at the grain surfaces.

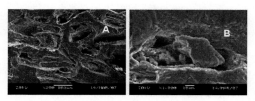

FIGURE 1. SEM micrograph revealing good connectivity (A). Al_2O_3 particles are clearly shown at the grain boundaries (B).

The resistivity measurement is presented in Fig. 2 in the temperature range 12-300K. The figure reveals an increase in the normal state resistivity as the concentration of Al_2O_3 nano-particles increases. It is interesting to notice that doping has very little effect on the transition temperature in all prepared samples.

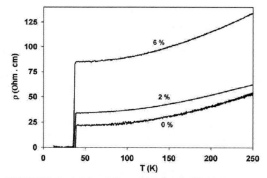

FIGURE 2. Resistivity of the samples doped with Al_2O_3 nano-particles.

Magnetization measurements are presented in Figs. 3 and 4. In Fig. 3, we present the hysteresis loops measured at 4.2K for all samples.

One can clearly observe two different behaviors. Below 2 Tesla, large unstable behavior is clearly seen in the figure, usually in both branches of the hysteresis loops. Above 2 Tesla, the loops follow the normal behavior commonly seen in hysteresis loops. This unstable behavior, commonly referred to as "vortex avalanche" has been observed in various high temperature superconducting samples with high critical current density.

Local heating causes tiny regions in the sample to turn normal, causing a sudden rush of the trapped flux to these areas. As a result a sudden drop in the magnetization is observed. Moreover, the remanent magnetization is reduced.

accordingly as can be seen in all hysteresis loops presented in Fig. 3. For all doped samples, the magnetization above 2 Tesla is higher than the magnetization of the pure sample, indicating higher trapped flux in doped samples. However, since no vortex avalanche is observed in the pure sample, the magnetization below 2 Tesla is higher than that of the doped samples. We have repeated the same measurements at higher temperature at which vortex avalanche completely disappear, namely at 20K. The results are presented in Fig. 4. Nearly all hysteresis loops for the doped samples are wider than the loops for the pure sample at all fields.

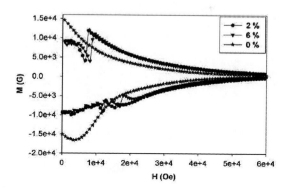

FIGURE 3. Hysteresis loops at 4.2K for MgB_2 samples doped with various concentrations of Al_2O_3 nano-particles. Vortex avalanche clearly shows below 2 Tesla for all doped samples.

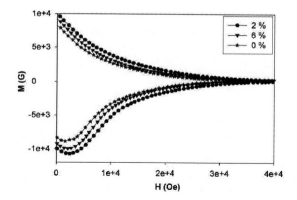

FIGURE 4. Hysteresis loops at 20K for MgB_2 samples doped with various concentrations of Al_2O_3 nano-particles. Vortex avalanche has completely vanished.

All hysteresis loops (Figs. 3 and 4) indicates wider loops in all doped samples: at low temperatures (<20K) above the vortex avalanches (>2 Tesla). At higher temperatures similar results can be seen nearly for all doped samples. This indicates that higher critical current density and higher flux trapping in nearly all doped samples. It is also noted that the irreversibility field has increased in all doped samples. The maximum increase was achieved in sample doped with 2% Al_2O_3, as shown in Fig. 5. It is interesting to observe that vortex instability is maximum in the highest doped samples (6% Al_2O_3), where the normal state resistivity is also maximum.

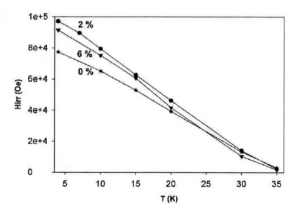

FIGURE 5. SEM micrograph revealing good connectivity (A). Al₂O₃ particles are clearly shown at the grain boundaries (B).

CONCLUSION

Magnetization measurements revealed a sizable increase in the width of the hysteresis loops in MgB_2 samples doped with Al_2O_3 nano-particles. We also observed an increase in vortex avalanche in doped samples that persists up to 20K in samples doped with 6% Al_2O_3. Hysteresis loops measurements also reveal significant increase in the irreversibility field (Hirr) as a result of doping.

ACKNOWLEDGMENTS

We would like to acknowledge the help and support of King Fahd University of Petroleum and Minerals. This work has been supported under fast track project No. FT-2004 -12.

REFERENCES

1. J. Nagamatsu, N. Nakagawa, T. Muranaka, Y. Zenitani, and J. Akimitsu, Nature **410**, 63 (2001).
2. Physica C **385**, Issues 1-2, (1 March 2003); special issue on MgB_2.
3. D. K. Finnemore, J. E. Ostenson, S. L. Bud'ko, G. Lapertot, P. C. Canfield, Phys. Rev. Lett. **86**, 2420 (2001).
4. J. S. Slusky, N. Rogado, K. W. Reagan, M. A. Hayward, P. Khalifah, T. He, K. Inumaru, S. Loureiro, M. K. Hass, H. W. Zandbergen, R. J. Cava, Nature **410**, 343 (2001).
5. K. P. Singh, Y, V. P. S. Awana; M. Shahabuddin, M. Husainy, R. B. Saxena, Rashmi Nigam, M. A. Ansari, Anurag Gupta, Himanshu Narayan, S. K. Halder, and H. Kishan, Modern Physics Letters B **20**, 1763 (2006).

Synthesis and Characterization of ZnO Nanostructures

M. Faiz and N. Tabet

Surface Science Laboratory, Physics Department, King Fahd University of Petroleum and Minerals,
Dhahran 31261, Saudi Arabia
mnfaiz@kfupm.edu.sa

Abstract. Nanocrystalline ZnO of various shapes was obtained without the use of catalysts by dry oxidation of Zn. Scanning Electron Microscopy results showed that the oxidation of zinc single crystals at temperatures below the melting point of Zn lead to the formation of a zinc oxide layer of flakes-type structure covering the surface of the sample. Above the melting point, long nanofibers were formed. The density and the length of these nanofibers increase as the temperature and the oxidation time increase. These results underline the importance of the role of the ZnO vapor pressure on the growth of nanocrystalline zinc oxide. X-ray Photoelectron Spectroscopy results confirmed the formation of oxide layer within a few minutes of heatinghe exposure of a Pt substrate to zinc foil during oxidation treatment led to the formation of long entangled nanofibers in the region of the substrate exposed directly to the zinc foil. High quality ZnO nanocrystals of hexagonal shape were obtained on the region of the substrate covered by a mica sheet. X-ray Diffraction results showed that the lattice parameters of the nanofibers are 1 % smaller than those of microcrystalline ZnO. Experiments are in progress to grow unidirectional nanofibers under high electric field.

Keywords: Nanostructure, Nanowire, Nanocrystals, XPS, Oxidation.
PACS: 68.37Hk, 81.07.-b

INTRODUCTION

The research interest in zinc oxide, a wide gap semiconductor, has increased drastically with the emergence of the field of nanomaterials. The growth of nanocrystals is of crucial importance for the development of electronic and optoelectronic nanodevices [1]. Nanotips can be used as probing tips for high resolution atomic force microscopy, as photonic crystals for waveguides and as field emitters for flat panel displays [2]. Laser emission has been reported from ZnO nanorods grown on sapphire substrate [1]. ZnO nanostructures have been synthesized by many groups using various techniques including catalyst-assisted Vapor Liquid Solid (VLS) [1,3-4] and thermal evaporation [5]. The development of synthesis methods that do not involve the use of catalysts is of particular interest as they reduce contamination and the cost. In this work, we report on the growth of ZnO nanostructures of various shapes obtained via dry oxidation of metallic Zn substrates.

EXPERIMENTAL

Zinc single crystals were mechanically polished perpendicular to the [0001] growth axis down to 5 microns size. Heat treatments of various durations were carried out at temperatures ranging from 300 °C to 600 °C under air. Oxidation of polycrystalline zinc foils of 1 mm thick was also investigated. SEM images were obtained using a scanning electron microscope (type JEOL-JSM6460). XRD experiments were carried out using a fully automated X-

CP929, *Nanotechnology and Its Applications, First Sharjah International Conference*
edited by Y. I. Salamin, N. M. Hamdan, H. Al-Awadhi, N. M. Jisrawi, and N. Tabet
© 2007 American Institute of Physics 978-0-7354-0439-7/07/$23.00

Ray Diffractometer (type Bruker XRD-D8). The incident photons were produced from a Cu anode (K_α, λ = 1.54056 $\overset{\circ}{A}$).

XPS spectra were recorded before and after the oxidation treatments. Photons from an aluminum anode (K_α, 1486.6 eV) were used as incident beam. The energy resolution of the Spectrometer (type VG-ESCALAB MKII) was about 1 eV. Zn 2p, Zn 3d, O 1s and C 1s lines were systematically recorded. The charge shift correction of the binding energies was done by using the C 1s line (E_b = 284.6 eV), stemming from the contamination layer, as a reference [6].

RESULTS AND DISCUSSION

Figure 1 (a, b) shows SEM images of the surface of three different Zn crystals after oxidation treatments at T = 300 °C for 19 hours and T = 400 °C for 4 days, respectively. One can notice in Figure 1a the presence of a pore that has the hexagonal symmetry of the (0001) oriented surface. The bright dots are ZnO particles. As the oxidation temperature and time increase, the surface becomes covered by an oxide layer of flakes-type structure.

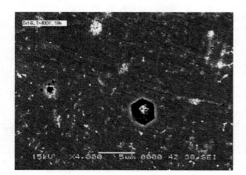 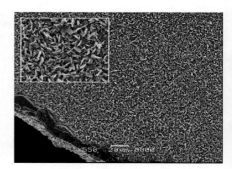

FIGURE 1. SEM images of Zn crystals oxidized at (a) 300 °C, 19 hours (b) 400 °C, 4 days.

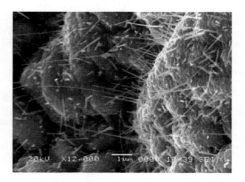

FIGURE 2. SEM Image of Zn Crystal oxidized at 450 °C for 15 minutes.

At temperatures exceeding the melting point of zinc (T_m = 419 °C), one notices the formation of fine fibers (whiskers) normal to the surface. Figure 2 shows the surface of a Zn crystal oxidized at T= 450 °C for 15 minutes. The fibers are of few microns length and about 100 nm diameter.

Figure 3 (a, b) shows the microstructure of the surface of Zn foil after oxidation at 500 °C for 6 hours and at 600 °C for 3 days, respectively. One can notice a significant increase of the density of ZnO fibers as the oxidation time and temperature increase.

FIGURE 3. SEM images of Zn polycrystalline foils oxidized at (a) 500 °C for 6 hours and (b) 600 °C for 3 days.

The formation of these nanofibers occurs probably through the evaporation of Zn atoms followed by an oxidation-deposition process on the surface of the sample. Therefore, the temperature is expected to influence drastically the growth process by modifying the vapor pressure of zinc in the atmosphere. The above SEM images show a clear evolution of the form of the oxide from small needle-like ZnO particles at 400 °C to long nanofibers of uniform cross section at 500 °C to large cone-shape needles at 600 °C.

The above results show that ZnO nanostructures are formed essentially at temperatures close to or above the Zinc melting point. As the growth experiments were carried out under air, one can assume that the surface of the sample is oxidized. ZnO vapor in the atmosphere is formed by thermal evaporation of zinc atoms and their subsequent oxidation. Therefore, the observed nanostructures were probably formed via Vapor Solid (VS) mechanism. It was reported in the mid nineteen fifties that heating metals under air can produce acicular oxides on the surface [7]. The anisotropic growth via the vapor solid mechanism is controlled by the temperature and the supersaturation of the atmosphere given the ratio α = p/p_0, p being the actual vapor pressure and p_0 the equilibrium pressure at the temperature of the experiment [8]. The increase of the density and the length of the fibers grown on zinc substrate as the temperature increased from 400 °C to 500 °C could be the result of the increase of the supersaturation ratio. Notice that the surface diffusion of adatoms to the top of the fibers can contribute to the growth mechanism [9]. An extensive nucleation on the side surface could block the surface diffusion of adatoms to the tip and lead to a thickening of the root part of the growing whiskers. This could explain the high density of cone-shaped needles observed after 3 day oxidation treatment of zinc foil at 600 °C (Figure 3b)..

XRD analysis of oxidized zinc foils revealed a complete oxidation of the foil only after 3 day treatment at 600 °C. A clear shift of the diffraction peaks towards higher Bragg angles was observed at 600 °C, suggesting a smaller volume of the unit cell of the nanofibers. The following values of the lattice parameters were obtained: a = 3.214 Å and c = 5.158 Å. These values are about 1 % smaller than those obtained for the samples oxidized at lower temperatures (450 °C and 500 °C) and those of commercial ZnO powder used as reference (a = 3.251 Å and c = 5.205 Å).

Analysis of the O 1s XPS spectra obtained from a Zn single crystal after successive oxidation treatments of short durations (from 10 to 60 minutes) showed the emergence of a new peak at 529.8 eV along with a shoulder at about

531.8 eV. The 529.8 eV peak is related to oxygen bound to Zinc atoms (Zn-O-Zn) while the shoulder has been assigned to the presence of water, as its binding energy lies between 531.5 eV (OH) and 533.0 eV (H_2O) [10,11] .

We have also grown ZnO nanofibers on mirror-polished Pt (001) substrate. The Pt crystal was placed on the top of zinc foil, the two samples being partially separated by mica sheets. The heat treatment was done under air at 500 °C for 12 hours. SEM images at higher magnification revealed the formation of high density of long entangled ZnO nanofiber on the area of Pt crystal that was exposed directly to the zinc foil (Figure 4a).

FIGURE 4. SEM images of entangled nanofibers and faceted nanocrystals on Pt substrate after heat treatment at 500 °C for 12 hours in presence of Zn foil.

The areas of the Pt substrate covered by mica showed the growth of a high density of ZnO nanocrystals of hexagon shape which indicates their excellent crystallinity (Figure 4b).Most of the nanocrystals had their growth direction [000 close to the normal to the substrate. This is in agreement with previous published results indicating a preferential grow of the ZnO nanofibers along the c-axis [3]. One can also notice, in Figure 4b, a preferential growth of the ZnO cryst along two polishing scratches indicating that the defects constitute preferential sites for the nucleation of ZnO crystals

CONCLUSION

Nanocrystalline ZnO of various forms was obtained by thermal oxidation of metallic zinc. The results show that the oxidation of zinc single crystals at temperatures below melting point of Zn lead to the formation of a zinc oxide layer of flakes-type structure covering the surface of the sample. Above the melting point, long nanofibers were formed. The density and the length of these nanofibers increase as the temperature and the oxidation time increase. The exposure of a Pt substrate to zinc foil during oxidation treatment led to the formation of long entangled nanofibers in the region of the substrate exposed to the zinc foil. Well faceted ZnO nanocrystals of hexagonal shape were grown on the region of the substrate covered by a mica sheet. These results underline the importance of the role of the ZnO vapor pressure on the growth of nanocrystalline zinc oxide.

ACKNOWLEDGMENT

The authors thank King Fahd University of Petroleum and Minerals for its support.

REFERENCES

1. M. H. Huang, S. Mao, H. Feick, H. Yan, Y. Wu, H. Kind, E. Weber, R. Russo and P. Yang, Science **292** (2001) 1897.
2. V. V. Poborchii, T. Tada and T. Kanayama, Appl. Phys. Lett. **75** (1999) 3276.

3. P. Yang, H. Yan, S. Mao, R. Russo, J. Johnson, R. Saykally, N. Morris, J. Pham, R. He and H. Choi, Adv. Funct. Mater. **12** (2002) 323.
4. M. H. Huang, Y. Wu, H. Feick, N. Tran, E. Weber and P.Yang, Adv. Mater. **13** (2001) 113.
5. T. Kim, T. Kawazoe, S. Yamazaki, M. Ohtsu and T. Sekigushi,   Appl. Phys. Lett. **84** (2004) 3358.
6. J. Moulder et al., Handbook of X-ray Electron Spectroscopy (Perkin-Elmer, 1992).
7. S. M. Arnold and S. E. Kounce, J. Appl. Phys. **27** (1956) 964.
8. J. M. Blakely and K. A. Jackson, J. Chem. Phys. **37** (1962) 428.
9. Y. Yin G. Zhang and Y. Xia, Adv. Funct. Mater. 12 (2002) 293.
10. Avalle, E. Santos, E. Leiva and V. A. Macagno, Thin Solids Films **219** (1992) 7.
11. L.J. Meng, C. P. Moreira de Sa and M. P dos Santos, Appl. Surf. Sci. **78** (1994) 57.

Modification in Synthesis of Anatase Titanium Oxide and Comparison of the Synthesis Methods

A. Hosseinnia, M. Pazouki, M. Kazemzad and M. Keyanpour-Rad

Energy Department, Materials and Energy Research Center, MeshkinDasht, Karaj, Iran
a-hosseinnia@merc.ac.ir

Abstract. In this research work, anatase titanium dioxide (TiO_2) was prepared by precipitation route using $TiCl_4$ in neutral pH. The dehydration of precipitate was performed by azeotropic extraction using benzene as solvent. After calcinations of product at 600 °C anatase form of titania was confirmed by XRD analysis. Precipitating $TiCl_4$ in pH=3 gives a product of mostly rutile and anatase. The sizes of titania nanoparticles obtained was less than 30 nm as determined by transmission electron microscopy (TEM) studies. In the other method of synthesis, titanium tetra isopropoxide was used in neutral pH. After calcinations of product at 600 °C, most of the titania obtained was brookite and anatase. Increasing the pH by addition of ammonia, $Ti(OH)_xCl_{4-x}$ will be formed first, and this increases the concentration of hydroxyl group in solution. In general when pH is higher, the amount of x in $Ti(OH)_xCl_{4-x}$ is more. FT-IR studies before calcination revealed that even in neutral pH this composition is present. Meanwhile the amount of x in the solution is highly related to the formation of anatase and rutile phases in the product. The higher the amount of x in the composition, the higher the amount of anatase phase present in the final product. The anatase obtained in addition to having interesting antibacterial properties, has other very interesting photocatalytic properties. It degraded methylene blue and rhodamie B in day light, which is rarely reported in the literature.

Keywords: titanium oxide, nanotechnology, photocatalyst, antibacterial.
PACS: 78.67.BF; 64.70.Nd; 61.46.HK

INTRODUCTION

TiO_2, a very important metal oxide, has been widely used in photovoltaic cells [1], photocatalysis [2], gas sensors [3], pigments [4], and so on. Nanocrystaline titanium dioxide particles have been intensively studied over the past two decades. Several different methods have been developed for the preparation of titania nanoparticles, which include the commonly used hydrolytic sol-gel process, non hydrolytic sol-gel process [5], hydrothermal and solvothermal methods [6].

Titanium dioxide is a wide band gap material ($E_g = 3.2$ eV) exhibiting photocatalytic decomposition and super hydrophilicity. Recently, photocatalysis based on titanium dioxide has attracted much attention in terms of environmental applications. Since the photocatalytic efficiency of TiO_2 increases by increasing its surface area, synthesis of the powders with high surface area is very important. Photocatalytic reactions proceed under UV-irradiation, with photon energy equal to or greater than the titanium band gap energy (for anatase: hv >3.2 e V, i. e. λ< 380 nm) which yields free photoelectrons (e^-) and holes (p^+). These photo-generated charge carriers are then able to interact through multi-step redox mechanisms with organic mater present at a TiO_2 particle surface [7, 8].

Titania has three polymorphs, anatase, brookite, and rutile. Rutile is the thermodynamically stable polymorph at all temperatures and pressures, while anatase and brookite are kinetic products. The phase change from anatase to rutile has been reported to occur in different temperature ranges from 600 to 1100 °C, depending on the preparation conditions for the anatase. It has been documented in the literature that hydrothermal processing of amorphous titania can result in the production of anatase, witch has been shown to have very high photo catalytic activity [9]. Since the anatase phase has a

CP929, *Nanotechnology and Its Applications, First Sharjah International Conference*
edited by Y. I. Salamin, N. M. Hamdan, H. Al-Awadhi, N. M. Jisrawi, and N. Tabet
© 2007 American Institute of Physics 978-0-7354-0439-7/07/$23.00

far higher photocatalytic activity than rutile phase, it is obviously desirable to stabilize the anatase phase as much as possible.

EXPERIMENTAL SECTION

Titania (TiO$_2$) was prepared according to the following three steps:
1. In a two necks Balloon, with a condenser and magnetic stirrer, 50 ml isopropanol was added while the solvent was stirred, 10 ml TiCl$_4$ was added to isopropanol. Then diluted ammonia (10%) was gradually dropped by a dropping funnel until pH 6. After 2 hrs precipitate was filtered and dried. The dehydration of precipitate was performed by azeotropic extraction using benzene as solvent. Then the product was calcinated at 600 °C in 4 hrs.
2. The above reaction was repeated but stopped in pH 3. After filtering of the precipitate, the treatments were followed similarly to the above method.
3. In this method (sol gel) in a beaker 15 ml Ti (IV) isopropoxide was added to 50 ml isopropanol, deionized water was sprayed on this solution while it was stirred by magnetic stirrer. The precipitate was filtered and dried. Then the powder was calcinated at 600 °C in 4 hrs.

The crystal structure, particle size, surface area of the prepared TiO$_2$ was characterized using x-ray diffraction (XRD), transmission electron microscopy (TEM) and infrared spectroscopy. Photocatalytic activity was evaluated by measurement of the decomposition of methylene blue and rhodamine B.

RESULTS AND DISCUSSION

X-Ray diffraction pattern of prepared TiO$_2$, from first method after dehydration shows only amorphous titania is made. Since in hydrolysis process TiCl$_4$ by ammonia solution a large amount of ammonium chloride is formed and with synthesized titania. In calcining this sample at 600 °C in 4 hrs all of the titania is transformed to anatase phase. Figure 1 shows the X-ray diffraction pattern of this sample. Therefore titania obtained from TiCl$_4$ precursor in pH 7 leaded to anatase phase formation and crystallization gets completed after heating the sample to 600 °C. The surface area of this sample measured by BET method was 71m^2/g. Particle sizes determination using Sherer formula from XRD pattern for this sample was 30 nm. From TEM images (Figure 2 and 3) it is clear that not only the particles are nearly regular but also their sizes are less than 30 nm, there are also some titania particles of around 10 nm in size.

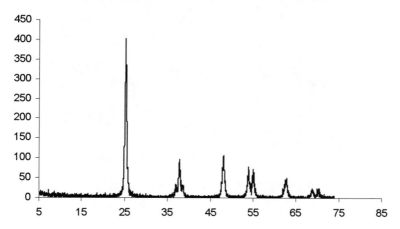

FIGURE 1. X-ray diffraction pattern of TiO$_2$ prepared TiCl$_4$ in pH 6-7.

153

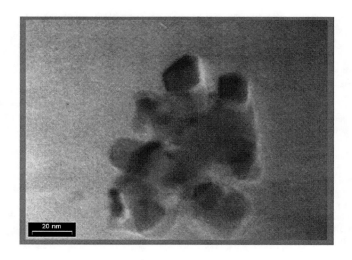

FIGURE 2. TEM image of TiO$_2$ from first method. The crystal size is smaller than 20 nm.

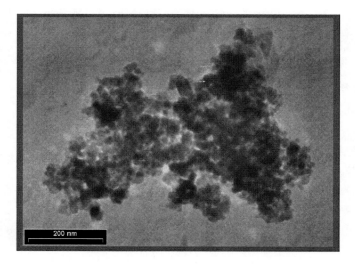

FIGURE 3. TEM image of TiO$_2$ from first method. The sizes are uniform.

In the second method which is similar to the first method with the difference of pH (pH=3). The product formed before calcination is amorphous, but in FTIR pattern there are some branched peaks which shows Ti(OH)$_x$Cl$_{4-x}$ is formed before dehydration. The value of x for initial product formed is proportional to the pH at which the synthesis was performed. X-ray diffraction of this sample (Figure 4) reveals that the product formed after calcination has 25% rutile and 75% anatase phase. This indicates that lowering the reaction pH will favor the rutile phase formation. We can ow confirm that the value of x in Ti(OH)$_x$Cl$_{4-x}$ will determine the whether the final product is anatase dominated or rutile dominated.

We also synthesized titania in a completely different (third) method. In this method the dehydration was not done with benzene, but rather spraying of water in the form of aerosol had resulted in the formation of fine particles. As shown by the XRD pattern (Fig. 5) the calcination temperature of 600 °C is not sufficient for the crystallization. The products formed are anatase and brookite and the majority of the product is in amorphous phase. The surface area of the calcined product was 250 m^2/g. The process of TiO2 in this method is progressed by Ti(OH)$_x$(isoC$_3$H$_7$O)$_{4-x}$ route and the value of x in pH 7 is close to 4 (from lowering of C-H stretching band of FTIR pattern). Breaking of organic bond in 600 °C occurs much higher than breaking of bond in Ti(OH)$_x$Cl$_{4-x}$. Therefore in using Ti(IV) isopropoxide transformation of Ti(OH)$_x$(isoC$_3$H$_7$O)$_{4-x}$ to TiO2 was fast and this has been prevented the crystallization of final product. In contrast transformation of Ti(OH)$_x$Cl$_{4-x}$ to TiO$_2$ was slower and the efficiency of the process in forming crystallized product is higher.

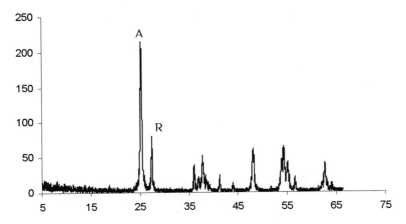

FIGURE 4. X-ray diffraction pattern of TiO$_2$ prepared TiCl$_4$ in pH 3 (A= Anatase, R= Rutile).

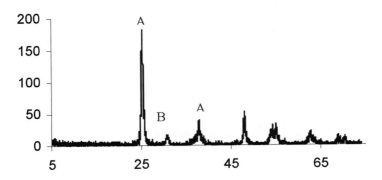

FIGURE 5. X-ray diffraction pattern of TiO$_2$ prepared Ti (isoC$_3$H$_7$O)$_4$ and water (A=Anatase, B=Brookite).

Photocatalytic activity of these samples with decomposition of methylene blue and rhodamine B was performed under sunlight and the results were compared. The decomposition capability of the products made by the first, second and third methods were half, 1 and a half hr, respectively. This indicates that although the third sample had a major amorphous phase but, due to the higher surface area of the anatase phase in this product, its activity is higher.

REFERENCES

1. B. O. Rregan, M. Gratzel, Nature **353**, 737 (1991).
2. J. C. Yu, W. Ho, L. Zhang, Chem. Commun. 1942 (2001).
3. N. Kumazawa, M. R. Islam, M. Takeuchi, J. Electroanal. Chem. **137**, 472 (1999).
4. C. Feldmann, Adv. Mater. **13**, 1301 (2001).
5. P. Arnal, R. J. D. Corvio, D. Leclercq, P. H. Mutin, A. Vioux, Chem. Mater. **9**, 694 (1997).
6. D. Pan, N Zhao, Q. Wang, Sh. Jiang, X. Ji, L. An, Adv. Mater. **17**, 1991 (2005).
7. Z. Zhang, C. C. Wang, R. Zakaria , J. Y. Ying, J. Phy. Chem. B **102**, 10871 (1998).
8. A. J. Maria, K. L. Yeung, C. Y. Lee, P. L. Yue, C. K. Chan, J. Catal. **192**, 185 (2000).
9. H. Kominami, T. Matsuura, K. Iwai, B. Ohtani, S. Nishimato, Y. Kera, Chem. Lett. **11**, 693 (1995).

Synthesis of Barium Titanate (BT) Nano Particles via Hydrothermal Route for the Production of BT-Polymer Nanocomposite

A. Habib[*], R. Haubner[†], G. Jakopic,[•] and N. Stelzer[*]

[*]Austrian Research Centers GmbH-ARC, Functional Materials, Seibersdorf, Austria
[†] Vienna University of Technology, Institute of Chemical Technologies and Analytics, Austria
[•]Joanneum Research, Weiz, Austria
amir.habib@arcs.ac.at

Abstract. Barium titanate (high-k dielectric material) nano-powders (approx. 30 nm to 60 nm) were synthesised using hydrothermal route under moderate conditions. Effect of temperature and time was studied using transmission electron microscopy (TEM), scanning electron microscopy (SEM) and X-ray diffraction techniques. Obtained barium titanate nano-powders were dispersed in thermoplastic polymethyl methacrylate (PMMA) to get homogeneous dispersions. Thin layers were obtained using these dispersions to achieve BaTiO₃ endorsed polymer layers by dip-coating for improved polymer insulators on various substrates e.g., glass, and Au sputtered silicon wafers. SEM and focused ion beam (FIB) techniques were used to study the dispersion of barium titanate nano-particles in PMMA. The layers obtained showed homogenous distribution of BaTiO₃ nano particles with no agglomeration.

Keywords: Hydrothermal synthesis, Dispersions, Polymer insulators, Agglomeration.
PACS: 82.35.Np

INTRODUCTION

Organic field effect transistors (oFET) are recently getting impressive improvement. These devices are attractive due to their light weight, cost effective production, and introduction of versatility in electronic devices like foldable electronic products [1]. For the improvement of oFET, one important goal is lowering the gate voltage. This can be achieved by replacing the currently used dielectrics by new high-k polymer composite dielectrics. Using materials with high dielectric constant enables the formation of equally thin layers as the currently used polymer dielectrics, but showing greater physical thickness.

BaTiO₃ (BT) perovskite is well known for its high dielectric constant. BT powder is often used in multilayered ceramic capacitors (MLCC) and BT layers are used as gate insulator. The synthesis by hydrothermal route is interesting as the sub-micrometer BT particles obtained via this route under moderate conditions, i.e., low reaction temperature and short reaction time, have exact stoichiometry, homogeneity, and uniform size distribution. Thermodynamics and kinetics of hydrothermal synthesis of barium titanate is investigated by many researchers [2- 6].

In order to synthesise a dielectric BaTiO₃-PMMA layer (polymer nano-composite layer -BT embedded in PMMA) offering good mechanical stability to the organic substrate at low processing temperature; a preparation route, consisting of a combination of the hydrothermal crystallization and the dispersion of these obtained nanoparticles in PMMA polymer matrix using surface active agents, is worked out.

The main challenge in achieving such a polymer nano-composite is mixing of incompatible inorganic BaTiO₃ and organic PMMA. Specific techniques to increase their compatibility are required for generating this inorganic-organic hybrid. Here we adopted the classical approach of using surfactants and block copolymers to stabilize BT in PMMA.

CP929, Nanotechnology and Its Applications, First Sharjah International Conference
edited by Y. I. Salamin, N. M. Hamdan, H. Al-Awadhi, N. M. Jisrawi, and N. Tabet
© 2007 American Institute of Physics 978-0-7354-0439-7/07/$23.00

EXPERIMENTAL

Hydrothermal Synthesis

Ba(OH)$_2$.8H$_2$O(from Sigma Aldrich) and TiO$_2$ P25 (degussa Germany approx. 25 nm; 30 % rutile, 70 % anatase) were used as precursors. The Ba to Ti ratio was kept at one. The precursors were added to Teflon vessel with bidistilled water. No mineraliser was used for pH adjustment. The pH of the solution at this concentration of Ba(OH)$_2$ was in the range 9-10. The minimum and maximum reaction times were 16 h and the time was 48 h respectively. The theoretical yield for the corresponding pH and purities of Ba(OH)$_2$ and TiO$_2$ precursors is 97 % [2]. The reactions were performed in an autoclave at autogenous pressure at 60, 90, 120 or 150°C while stirring.

The resulting slurry was subsequently washed with 1 M formic acid and bidistilled water to remove BaCO$_3$. The slurry was dried overnight in a furnace at the same temperature as the preceding reaction was performed. The dry cake was grinded to powder.

Powder samples were characterized using XRD, TEM and SEM techniques.

Dispersion of BT Nanoparticles in PMMA

Due to its uniform size distribution and better dispersion properties the sample synthesised at 90°C for 48 h was selected to obtain BT-PMMA composite.

Tetrahydrofuran (THF) and surface active agents as well as ultrasonic treatment were used to disperse the BT powders and afterwards PMMA-THF solution was added and refluxed below 100°C for 18 h. Several surface active agents were tested to reach best dispersion.

These dispersions were used for dip coating the glass and Au sputtered Si-wafer substrate. The layers were obtained at different dip coating speeds. These layers were characterised by using SEM and focused ion beam (FIB) techniques to see the distribution on BT in PMMA matrix.

RESULTS AND DISCUSSION

Hydrothermal Synthesis of BT

The perosvkite structure was observed for samples of BT obtained at 60°C for 48 h reaction with traces of unreacted titania by XRD anaylsis. This sample showed no sharp edges, amorphous and crystalline portion when observed under high resolution transmission electron microscopy (HRTEM) indicating an incomplete reaction (Figure 1). Energy filtered TEM (EFTEM) images for the same sample helped to observe unreacted titania at inner portion of the particles. In Figure 1 bright and dark contrast for titanium and oxygen, shows the presence TiO$_2$ at the core, while the barium was missing in that region. It can be concluded that barium has not yet diffused to this region as proposed by in-situ transformation reaction mechanism by Hertl [3] and the crystallization of BT has started as low as 60°C [7].

The reaction samples at 90, 120 and 150°C for different times showed improved crystallinity of BT at lower reaction times for higher reaction temperatures by XRD analysis. The morphology of the BT powders changed from semi crystalline with no sharp edges at 60°C to porous, cubic, and round shape at 90°C and 120°C, then changing to compact hexagonal particles at 150°C. The crystallite sizes for the samples were calculated using Scherrer's equation. A closer look at time dependence of crystallite size of BT reveals that higher reaction temperature results in small crystallite size, while no trend can be predicted regarding the effect of time on crystallite size of BT (Figure 2). The phase identified for all obtained BT powders was cubic.

TEM observation of BT synthesis at 90°C for 48 h showed that some particles had internal pores (Figure 3). Formation of pores in hydrothermally synthesised powders were attributed to outward transport of fast moving Ti-OH$_2^+$ or HTiO$_3^-$ through the oxide layer and a balancing inward flow of vacancies to vicinity of the TiO$_2$ interface [8].

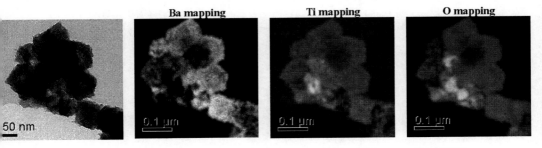

Ba mapping   **Ti mapping**   **O mapping**

50 nm   0.1 μm   0.1 μm   0.1 μm

FIGURE 1. Element distribution in BT powders synthesized at 60°C for 48 h by energy filtered TEM.

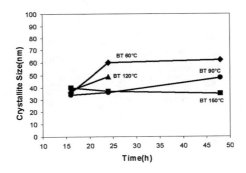

FIGURE 2. Time dependence of crystallite size during hydrothermal synthesis.

The particle size was found to be uniform and elemental distribution in the particles is homogenous with no signs of unreacted titania for these samples. Synthesis at higher temperature (150°C for 48 h), yielded nice compact particles with high crystallinity and hexagonal shape were observed (Figure 3).

SEM micrographs of obtained BT showed ultrafine agglomerated nanoparticles with uniform size distribution (Figure 4). The particles size was in the range of 30 nm to 60 nm. The internal pores were seen in samples, while hexagonal shape of the compact barium titanate was evident in the sample at 150°C for 24 h.

Dispersion of BaTiO$_3$ in PMMA

Usually two features of surfactants, i.e., adsorption at interface and self accumulation in supramolecular structure helps to stabilize the colloidal suspensions. The adsorption of surfactants onto inorganic particles depends on the chemical characteristics of the particles. Here, the BT particles obtained by hydrothermal synthesis are saturated with the hydroxyl groups (–OH) due to hydrothermal route. These hydroxyl groups can link to the polar head of the surfactants while the hydrophobic tail is sticking out to be suspended in organic solvent, which will further help to join these surface functionalized BT nanoparticles to the polymer matrix (PMMA). The surfactants can easily join the porous BT powders (i.e., sample BT 90°C 48 h).

The excess of surfactant in the PMMA helps the particles to stay disaggregated. The self assembly of BT particle within PMMA allows obtaining very thin layers with well dispersed BT at nanoscale in these thin films (Figure 5).

The SEM image of FIB cross section (Figure 5c) reveals the BT is distributed within the layer homogenously.

BT 90°C 48 h   **BT 150°C 48 h**

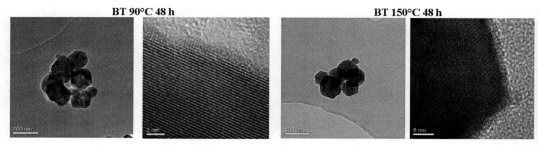

FIGURE 3. TEM images of BT powders showing pores and crystallinity at high synthesis temperatures

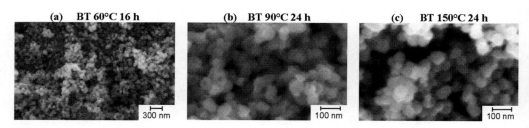

FIGURE 4. SEM images of BT samples (a) BT60°C 16 h, (b) BT90°C 24 h; pores in particles are evident and (c) BT150°C 24 h compact hexagonal shape started appearing.

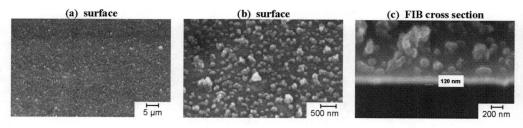

FIGURE 5. Surface and FIB-cross section of BT-PMMA layer showing homogenous distribution of BT nanoparticles.

CONCLUSION

The particle size of obtained BT is comparable to particle size of Ti precursor. This supports the in situ transformation mechanism proposed by W. Hertl [3] for the transformation of BT; which involves diffusion of Ba ions into the undissolved TiO_2 oxide, resulting in a shell of BT with unreacted TiO_2 core. The EFTEM images of sample BT shows that the outer surface has homogenous distribution of barium, titanium and oxygen whereas, the core has un-reacted titania.

The crystallite size of BT obtained is larger for lower reaction temperatures and no trends can be predicted about the particle size evolution by increasing the reaction time. The morphology of BT particles changes with temperature.

The hydroxyl groups (-OH) groups of BT obtained by hydrothermal synthesis helped to bond surfactants to porous surface and results in better dispersion of these surface functionalized BT in PMMA.

ACKNOWLEDGMENTS

We would like to thank Higher Education Commission of Pakistan for extending scholarship for PhD work of Mr. Amir Habib and to ISOTEC project for financial assistance.

REFERENCES

1. C. Reese, M. Roberts, Mang-mang Ling, and Zhenan Bao, *Materialstoday*, 20-27 (Sept. 2004).
2. M. M. Lencka and R.E. Raman, *Chem. Mater.* 5, 61-70 (1993).
3. W. Hertl, J. Am. Ceram. Soc. 71, 879-83 (1988).
4. S. Kaneko and F. Imoto, *Nippon kagaku Kaishi* 6, 985-90 (1975).
5. N. A. Ovramenko, L.I. Shvets, F.D. Ovcharenko, and B.Y. Kornilovich, *Dokl. Akad. Nauk SSSR*, 248, 889-91 (1979).
6. J. O. Eckert Jr., C.C. Hung-Houston, B.L. Gersten, M.M. Lencka and R.E. Riman, J. Am. Ceram. Soc. 79, 2929-39 (1996).
7. J. Moon, E. Suvaci, T. Li, S. Constantino, and J. H. Adhair, *J. Eur. Ceram. Soc.* 22, 809-815 (2002).
8. Y. Wang, H. Xu, X. Wang, X. Zhang, H. Jia, L. Zhang, and J. Qui, *J. Phys.Chem. B* 110, 13835-40 (2006).

Synthesis of Nano-Sized Nd$_2$O$_3$ Crystallites by Modified Co-Precipitation Methods

R. Malekfar, S. Arabgari, K. Motamedi, B. Parvin and M. Farhadi

Physics Department, Faculty of Basic Sciences, Tarbiat Modarres University, P.O. Box 14115-175, Tehran, Iran
Malekfar@Modares.ac.ir

Abstract. In this paper we present our extensive investigations on the synthesis of nano-sized Nd$_2$O$_3$ crystallites by different co-precipitation methods which include the standard co-precipitation and three modified co-precipitation techniques. The first modification was near zero centigrade co-precipitation. So we add the precipitator (NH$_4$HCO$_3$) to the nitrate solution while it was at 0°C of temperature. In the other method we performed the standard co-precipitation procedure vice versa. So we added the nitrate solution of neodymium to the NH$_4$HCO$_3$. In the fourth experiment we performed and used the above mentioned modifications together. So we added the nitrate solution to the precipitator while the whole system was kept at 0 °C temperature. The Raman spectra indicate that powders produced by reverse co-precipitation have better crystallite quality toward the three other methods. Also SEM images reveal that all modification technique has significant effect on the agglomeration of the powders which is another important factor of the quality of the nano powders. Finally XRD Patterns show that in all the different kind of co-precipitation methods pure neodymium oxide phase are produced. It is also confirmed that the sizes of the synthesized powders by our reverse technique are finer than the powders synthesized at in the near zero centigrade by the standard co-precipitation method.

Keywords: *Nano-sized, Nd$_2$O$_3$, co-precipitation, Raman spectroscopy, FTIR*
PACS: 78.67.-n, 78.67.Bf, 75.50.Tt

INTRODUCTION

Among the different examined methods co-precipitation technique of synthesizing nano-sized powders is one of the most effective techniques [1]. Our group for the first time successfully suggested and performed near zero centigrade temperature co-precipitation (ice) method [2] and then ice-reversed method was examined in this research. These four methods are different ways of performing the co-precipitation technique that have been examined and the results are compared in this article.

EXPERIMENTAL

In this research we performed the following four techniques in order to synthesize Nd$_2$O$_3$ nanoparticles:

1. The steps of the standard co-precipitation technique of producing nano-sized neodymium oxide powder are based on dissolving the ordinary neodymium oxide in watery nitric acid to obtain its nitrate solution. We used NH$_4$HCO$_3$ as precipitator with concentration of 3M. In the diluted NH$_4$HCO$_3$ decomposes to two compounds and only one of them participates in the sedimentation process. Nd(OH)$_3$ sediments at a certain pH which is calculable from the solubility index k$_{sp}$ of the reaction factors. The above reaction happens slowly, so during the precipitator adding process into the nitrate the pH increased very rapidly, but gradually after hours, it

CP929, *Nanotechnology and Its Applications, First Sharjah International Conference*
edited by Y. I. Salamin, N. M. Hamdan, H. Al-Awadhi, N. M. Jisrawi, and N. Tabet
© 2007 American Institute of Physics 978-0-7354-0439-7/07/$23.00

started to decrease and the real pH number of the colloidal solution was indicated. Now we had the colloidal solution whose sediment must be separated from the undesirable compounds of NH_3^+ and NO_3. So we filtered and washed the colloidal solution by de-ionized water several times. Then we dried the hydroxide sedimentation by heating it at 100°C for 30 minutes. The last stage was calcination. We had two calcination steps. The first step was performed at 700°C twice, in which we used mechanical milling after each step. The second step of calcination was carried out at high temperature (over 1000°C). The powders were heated twice, followed by ball milling every time. Every process lasted for 5 hours. After calcination nano-sized powders were synthesized.

2. The modified co-precipitation technique which we call "the near zero degree centigrade co-precipitation technique" we believe has been suggested and performed by our group for the first time. This technique can be carried out by performing the sedimentation process at 0°C temperature. This modification is executed by a change in the second stage. So we added the precipitator (NH_4HCO_3) at room temperature to the nitrate solution which was kept at 0°C temperature. The other stages in this modified technique are exactly the same as in standard co-precipitation.

3. In the reversed co-precipitation technique we performed the standard co-precipitation procedure in reverse. This modification is performed by a change in the second stage of standard co-precipitation. Thus we added the nitrate solution to the precipitator (NH_4HCO_3) drop-wise and after obtaining the colloidal solution we executed the other stages as before.

4. The near zero degree centigrade reversed co-precipitation technique can be performed by applying the last two methods conditions simultaneously. So in this method at the second stage we kept the precipitator (NH_4HCO_3) near zero centigrade temperature and added the nitrate solution drop wise to it.

In Table 1, the whole processes which were carried out for synthesizing 8 different samples are presented. Now our task is to compare the grain sizes, agglomeration and molecular structures of the neodymium oxide powders produced in standard co-precipitation and our suggested modified co-precipitation methods.

RESULTS AND DISCUSSIONS

One of the best methods of calculating the grain sizes and indicating their agglomeration is by using their SEM images, which is a very direct investigation. In Fig. 1 the SEM images of the 6 samples are shown.

As shown clearly in the SEM images, the powders produced by different methods have variable characterizations. In *standard co-precipitation method* the grains are too agglomerate. There is a clear difficulty to measure the sizes of grains.

In the *near zero centigrade method* the grains are less agglomerated. Their sizes look quite more suitable.

In the *near zero centigrade-reverse technique* the grains are not agglomerated and they seem to be finer than the grains of previous methods. It is a bit hard to compare their sizes, thus it should be checked by other techniques, for example by XRD patterns, which will be carried out in the next section.

Since co-precipitation is a well-known method of synthesizing a wide range of nano-sized powders, the modified methods successfully suggest modifications that can replace the standard co-precipitation method. The reasons of the decrease in the agglomeration can be explained by the precipitation theory.

The process of precipitation (sedimentation) could be expressed with regard to two sedimentary processes, i.e., nucleation and particle growth. The size of particles of sedimentation depends on the extent of predominating effect of one of these two processes. The nucleation is a process in which the minimum number of ions or molecules can be united in order to form the second stable phase. The precipitation can be continued either by forming new nucleation or by setting ions on existing nucleation. If the former predominates, sedimentation with a number of particles will be produced. If the growth of particles dominates, the macro-particles, with the minimum possible number, will be produced. So we predict if the sedimentation process happens at low temperature, the formed sedimentary nuclei will have low kinetic energy and with much less touches of each other in order to produce bigger particles. This issue causes our desirable process, the nucleation, predominates the particle growth.

Different experiments were performed and used along with modifications which include the above mentioned cases. However, when we added the nitrate solution to the precipitator, this step stopped to increase the nucleation process.

163

TABLE 1. The Samples Synthesized and Their Characterization.

Method	Calcinations temperatures	
Standard co-precipitation	900°C	1100°C
Near zero centigrade co-precipitation	900°C	1100°C
Reverse co-precipitation	900°C	1100°C
Ice-reverse co-precipitation	900°C	1100°C

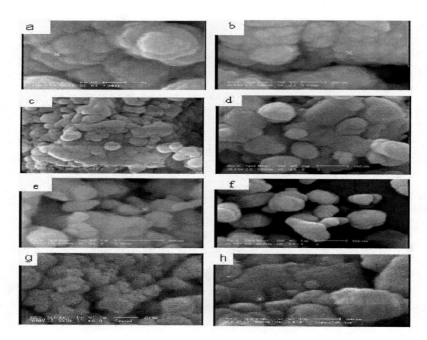

FIGURE 1. The SEM images of the samples produced at: a) 900°C standard co-precipitation, b) 1100°C standard co-precipitation, 900°C ice co-precipitation. d) 1100°C ice co-precipitation, e) 900°C ice-reverse co-precipitation, f) 1100°C ice-reverse precipitation, g) 900°C reverse co-precipitation, h) 1100°C reverse co-precipitation.

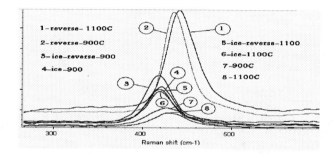

FIGURE2. Raman Spectra of the samples synthesized by different modification of co-precipitation technique.

Crystallization and Grain Size

By investigating the XRD patterns of the samples the grain sizes were calculated by using the Debye- Sherrer formula. The results of the calculation are listed in Table 2.

TABLE 2. Grain Size of Samples measured from XRD patterns.

Method	Calcinations temperature (°C)	Grain size (nm)
Standard	900	41
Near zero centigrade	900	40
Near zero centigrade	1100	22
Reverse	900	29
Reverse	1100	25
Near zero centigrade- reverse	900	30
Near zero centigrade- reverse	1100	25

As the XRD results show, the calcinations temperature has a great effect on the grain size, but for the samples made at the same calcinations temperature, the samples produced in reverse technique are finer than the powders synthesized only at near zero centigrade co-precipitation method.

Back scattering Raman spectroscopy reveals the vibrational and rotational frequencies. It can be used to study the frequency shift in identifying the peaks of the crystals due to better crystallite qualification. The results are shown in the Fig. 2.

It has been found in our studies that all the powder samples produced by reverse co-precipitation have better quality [3,4] than powder samples synthesized by ice methods and the lowest quality belongs to the nano-crystallite synthesized by standard co-precipitation. There is also a tendency of shifting in some Raman bands. This shifting is more adapted to the SEM images. As the agglomeration and grain sizes start to decrease the peaks of the crystal shift to higher frequencies.

We have also used near normal FT-IR reflection spectroscopy for calculating the dispersive parameters such as the refractive index of the synthesized nano powders which were compacted to disc shape pellets. The pellets in this stage are suitable for near normal reflection FTIR spectroscopy in order to record their reflection spectra. According to the Kramers-Kronig relations analysis of the near normal FTIR reflection spectroscopy, we can calculate the dispersion relations, e.g., the real and imaginary refractive indices, n and k, and also the dielectric indices, e_1 and e_2 and from the near normal reflection spectra. The obtained results are in reasonable agreement with the information collected from Raman spectra, XRD patterns and SEM images.

CONCLUSION

Applying different modifications of co-precipitation methods give various conclusions in synthesizing nanc powders. Standard co-precipitation gives intense agglomerated grains. Near zero co-precipitation method gives finer grains and reverse co-precipitation method helps to produce crystallite with better structure.

REFERENCES

1. A. K. Pradhan, Kai Zhang, G. B. Loutts, Materials Research Bulletin **39** (2004) 1291–1298.
2. R. Malekfar, S. Arabgari, B. Parvin and K. Motamedi, 2nd International Spectroscopy Conference (March 25-28, 2007) Sousse – Tunisia.
3. R. Malekfar, S. Arabgari, K. Motamedi and B. Parvin, NANOSS4 Workshop (17-21 September 2006) Rathen, Germany.
4. R. Malekfar, S. Arabgari, B. Parvin and K. Motamedi, 13th Iranian Conference on Optics and Photonics, ICOP2007 (6-8 February 2007) ITRC, Tehran, Iran.

Micro-Raman Scattering Spectroscopy of Nanocrystalline Phases of TiO₂ Prepared by Sol-Gel Method

R. Malekfar[1], M. Mozaffari[2] and S. Mihanyar[2]

[1] *Physics Department, Faculty of Basic Sciences, Tarbiat Modarres University, P. O. Box 14115-175, Tehran, Iran*
Malekfar@Modares.ac.ir
[2] *Physics Department, Faculty of Science, Islamic Azad University, North Tehran Branch, Tehran, Iran*

Abstract. In this paper the synthesis of the fine and homogenous nano crystalline TiO₂ particles obtained from Ti (OBU)₄ in the presence of H₂O₂ by Sol-Gel technique is investigated. The calcinations of the precursor powders of TiO₂ samples were performed at 550°C, 750°C and 950°C. The metastable anatase form can be transformed thermally into rutile phase but the rate critically depends on the specific growth environment. The collected synthesized powders were characterized by micro Raman back-scattering spectroscopy, FTIR Spectroscopy, XRD and SEM imaging. With XRD results and also from the sample surface morphology, the size of nanocrystalline particles were determined, which showed that our samples had nano scale and also confirmed presence of pure TiO₂ phases in the synthesized structures. The recorded Raman spectra confirmed that nanocrystalline anatase phase powder, which is famous for its application as photo catalyst, was produced at 550°C. At 750 °C and 950 °C mixed and pure rutile phases were observed, respectively.

Keywords: TiO₂ nano-particles, anatase, rutile, Raman spectroscopy.
PACS: 78.67.-n, 78.67.Bf, 75.50.

INTRODUCTION

Titanium dioxide has three phases in nature, rutile, anatase, and brookite. Rutile has high reflection index, visible transparency and ultraviolet absorption properties. Thus it has been widely applied in pigments and opacifiers. Anatase is chemically and optically active, suitable for catalyst support and as a photo electrochemical material [1-4]. Brookite is rare in nature, but recently it has been found that it is more electrochemically active than anatase and is a good candidate for photovoltaic devices [4-5]. Therefore, it is interesting to investigate the phase stability and phase transition of TiO₂ crystals.

Conventionally, preparation of nanosized TiO₂ needs heat treatment. The heating process causes the grain growth and thermal stability, which prohibits the production of nanosized TiO₂ powder. In bulk material form, rutile is stable, whereas brookite and anatase are metastable and easy to transform to rutile when heated. The transformation temperature varies according to particle size, impurity level, and atmospheric condition [6].

It is well known that the particle size is one of the factors to influence the photocatalytic performance of TiO₂. So in this paper, we report the preparation of samples of TiO₂ nanocrystals with grain sizes of anatase and rutile. We study the influence of the transition by the grain size, grain shape and phase contents change by heat treatment. Many methods have been employed to control size and morphology of TiO₂, titania, such as sol-gel, hydrothermal technique and co-precipitation. Based on synthesis of the anatase TiO₂ nanocrystals prepared by sol-gel process of tetra butyl titanate, we investigate their Raman spectra of the nano-sized powders prepared at various annealing temperatures and particle sizes in order to get a better understanding of the lattice vibration characterization in ultra fine particle TiO₂ systems. Anatase

CP929, *Nanotechnology and Its Applications, First Sharjah International Conference*
edited by Y. I. Salamin, N. M. Hamdan, H. Al-Awadhi, N. M. Jisrawi, and N. Tabet
© 2007 American Institute of Physics 978-0-7354-0439-7/07/$23.00

to rutile structural transformation was studied by X-ray diffraction, back-scattering micro Raman spectroscopy and SEM imaging.

EXPERIMENTAL

The precursor solutions were prepared in a rather specific procedure by adding 10ml Ti(OBU)$_4$ drop-wise into ice-cold H$_2$O$_2$ under stirring conditions. In order to prevent probable explosion, all the experiments were carried out in an ice bath. After adding a few drops, the colorless H$_2$O$_2$ solution became yellow then a red solution was obtained which was shortly followed by a strong exothermic reaction due to the unstable nature of H$_2$O$_2$ and Ti compounds. Ending Ti(OBU)$_4$ is companion with giving off H$_2$O$_2$,O$_2$ and butyl alcohol. Finally a yellow transparent Peroxo-Poly Titanic (PPT) acid gel was prepared. This gel was kept at room temperature for 16h in order to lose its water, before drying at 150°C for 5h in an oven. At this stage PPT was transformed into an amorphous TiO$_2$ powder. The produced yellow powder was grinded for 1 hour. The calcination precursors of the TiO$_2$ samples were performed at 250°C, 350°C and 450°C for an hour with ramping up speed of 5°C/min heating rate. Then they were kept at 550°C, 750°C and 950°C for 30 min at each stage. The temperature increase rate was 2°C/min. Finally white, smooth and homogenous TiO$_2$ nanoparticles were observed.

RESULTS AND DISCUSSION

Back-scattering micro-Raman scattering was employed to investigate the evaluation of anatase and rutile phases in synthesized TiO$_2$ nano-crystallites. Raman features of TiO$_2$ nanocrystals could be identified by those of the bulk TiO$_2$ and phonon confinement, non-stoichiometry and surface pressure are responsible for the blue shift and broadening of the lowest frequency E$_g$ Raman mode [15].

It was clear that gradually by increasing calcinations temperature at 350°C anatase phase of TiO$_2$ nano powders starts to be formed. By investigating the Raman spectra it is obvious that there are three main distinctive Raman modes: B$_{1g}$, A$_{1g}$ and E$_g$. They were at 378cm^{-1}, 496 cm^{-1} and 617 cm^{-1}, respectively. By increasing the temperature and up to 550°C anatase phase was maintained. By increasing the temperature to 750°C ordered crystal structure can be formed and gradually anatase phase was changed to rutile phase. At this temperature the Raman spectra were a mixture of anatase and turile phases. However, at 950°C pure rutile phase was observed. Its two modes were located at 427cm^{-1} and 590cm^{-1}. As the calcinations temperature increased, the structure became more regular and the grain particle sizes larger, and also the Raman peaks get border. There is also a dramatic increase in the intensity of the lowest frequency E$_g$ mode, [15]. Figure.1 also exhibits the Raman spectra of the nanocrystallite powders of TiO2 prepared at different calcinations temperature. The transformation from anatase to rutile (a→r transformation) phase started at 550°C for TiO$_2$ powders prepared from Ti(OBu)$_4$ and H$_2$O$_2$ by sol-gel method.

Figure.2 shows the XRD pattern of TiO$_2$ powders calcinated at 550°C, 750°C and 950°C. It is shown that by increasing calcinations temperature the width of Raman peaks start to decrease and also their intensity become more intense. Growth of grains and formation of crystalline structure appeared on the particle surfaces by increasing the calcinations temperature.

When the powders annealed at 550°C, the samples showed clear, broadened anatase peaks. However, calcinations at 750°C, a mixed of anatase and rutile phases were detected. Finally, when calcinations were performed at 950°C, pure rutile phase was obtained. By increasing the annealing temperature the anatase structure starts to appear and was then progressively can be transformed to a rutile structure phase.

Powder XRD patterns also were used for the phase identification and also for the determination of crystallite size and the estimation of the quantity ratio of anatase to rutile phases. The fraction of rutile phase in each sample was determined via the generally accepted method. This method is consisted to the measurement of the mass fraction of rutile (χ_R) and was determined from the following expression [12, 13].

$$\chi_R = \frac{1}{1+0.8\left(\dfrac{I_A}{I_R}\right)}, \qquad (1)$$

in which I_A and I_R are the integrated line intensities of the anatase (101) and rutile (110), respectively. For the prepared sample at 550°C, I_A and I_R are 100 and 9.1, respectively. So by using expression (1) we concluded that the obtained structure is made of anatase phase and only 10% of the structure belongs to the rutile phase. At 750°C rutile to anatase ratio phase is 97%. The XRD pattern of the sample calcinated at 950°C consists of very sharp peaks of pure rutile and the peaks due to anatase have completely disappeared.

The XRD measurement also can be used for identifying the nanocrystallite grain sizes of TiO_2 powders according to the Debey-Scherrer equation, [14].

$$D = \frac{K\lambda}{\beta \cos\theta},$$ (2)

where λ is the X-ray wavelength which in our experiment was equal to 0.154 nm, β, the full width at half-maximum (FWHM), θ, the half diffraction angle, and k is usually taken as 0.89. Expression (2) determined that at 550°C for the prepared anatase phase and at 750°C for rutile phase, particle sizes are 21.9nm and 90nm.

The SEM image of the TiO_2 nanopowder and the corresponding grain size distribution are presented in fig.3. As can be seen from the figures, most of the particles in the first three samples have spherical shape. However, sample (d) has a different shape distribution in compare with the previous three samples, and clearly it has a cubic grain shape, which is due to the structural change. The sizes of the nanoparticle powders in the above mentioned range of temperatures were estimated to be from 57nm to 78nm. Our results were in good agreement with the results obtained from the Debey-Scherrer formula. Thus while the calcinations temperatures increase, the grain sizes start to increase.

In order to find the optical coefficient of nano-particle TiO_2 FTIR reflection spectroscopy was employed. Figure 4 shows that each of the anatase and rutile phases have three main modes, one A_{2u} and two E_{2u} modes ($A_{2u} + E_{2u}$), at the spectral rang of 400 cm^{-1} up to 1000 cm^{-1}. For anatase and rutile phases, these main modes are A_{2u} at 464 cm^{-1}, E_u at 414 cm^{-1} and E_u at 400 cm^{-1}.

The positions of the main peaks for the IR spectra of the two phases are approximately equal, while the intensities are different. There are three IR active modes ($A_{2u}+2E_u$) and also Raman modes of E_g type which originate from oxygen vibration. The additional modes can be understood in terms of IR-forbidden modes, since the oscillator frequencies are very similar to the two strongest E_g modes in the Raman spectrum of anatase TiO_2.

Actually, the existence of the small amount of Ti3$^+$ substituted on Ti^{4+} sites locally may remove the neutrality of changes in E_g vibration modes, which may cause a slight lattice distortion and appearance of IR-forbidden modes. The Ti^{3+} states in the TiO_2 lattice originate from nonstoichiometery, oxygen deficiency and ion intercalation, as well as a surface adsorbed species and other surface or interface states [9, 10].

The optical and dielectric parameters of the nanoparticles can be calculated by using the near normal reflection spectra and applying the Kramers- Kronig relations. Accordingly the real part of the refractive index for both anatase and rutile phases of TiO_2 nanoparticles in mid infrared region have been found which are around 2.25 for both phases.

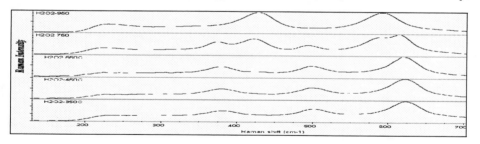

FIGURE 1. The effects of calcinations temperatures on the Raman spectra for five nanocrystallite samples produced by the sol-gel method.

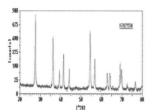 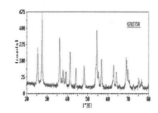 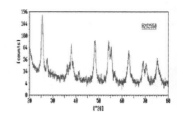

FIGURE 2. XRD patterns of TiO₂ powders calcinated at: a) 550°C, b) 750°C and c) 950°C for 30minuts.

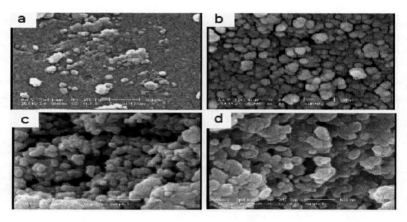

FIGURE 3. Typical SEM images of the synthesized samples. Increasing of the grain growths of TiO₂ nanosized crystallites at (a) 150°C, (b) 550°C, (c) 750°C and (d) 950°C temperatures.

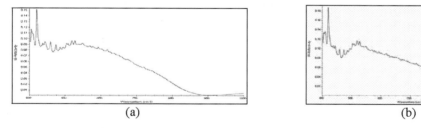

(a)   (b)

FIGURE 4. Near normal IR reflection spectra of nanocrystalline powders of TiO₂: (a) in anatase phase and (b) in rutile phases prepared at 400°C and 1000°C, respectively.

CONCULOSIONS

To conclude, the main features of the investigation presented above may be summarized as follows. Raman spectroscopy showed that as the calcinations temperature increased, the structure became more regular and the grain particle sizes larger, and also the Raman peaks get broader. XRD patterns indicated that by increasing calcinations temperature the width of Raman peaks start to decrease and also their intensity become more intense.

Growth of grains and formation of crystalline structure appeared on the particle surfaces by increasing the calcinations temperature. The sizes of the nanoparticle powders were estimated to be from 57 nm to 78 nm by SEM imaging, thus while the calcinations temperatures increase, the grain sizes start to increase. FTIR studies confirmed that the real part of the refractive index for both anatase and rutile phases of TiO_2 nanoparticles in the mid infrared region have been found which are around 2.25 for both phases.

REFERENCES

1. B. O. Regan, M.Gratzel, Nature **353** (1991) 737-739.
2. J. C. Yu, Li. Z. Zhang et al., Chem. Mater. **14** (2002) 4647-4653.
3. B. Xia, H. Hung et al., Mater. Sci. Eng. B (1999) 150-154.
4. K. R. Zhu, M. S. Zhang et al, Mater. Sci. Eng. A **403** (2005) 87-93.
5. M. Koelsch, S.Cassaignon et al., Thin Solid Films **403/404** (2000) 310-312.
6. P. P. Ahonen, J. Joutsensaari et al., J. Aerosol Sci. **32** (2001) 615-630.
7. K. Jezierski, W. Misiewicz et al., Optica Aplicata, Vol. XI, No. 4 (1981).
8. X. Wang, C. Neff et al., Advanced Materials **17** (2005) 2103-2106.
9. A. M. Eppler, I. M. Ballard et al., Physica E **14** (2002).
10. M. G. Brojcin, M. J. Scepanvic et al., Journal of Physics D: Applied Physics **38** (2005) 1415-1420.
11. P. M. Kumar, S. Badrinarayanan et al., Thin Solid Films **358** (2000) 122-130.
12. J. A. Gamboa, D. M. Pasquevich et al., J. Am. Ceram. Soc. **75** (1992) 2934-2938.
13. M. P. Zheng, M. Y. Gu et al., Mat. Sci. Eng. B **87** (2001) 197-201.
14. X. Liu, X. Wang et al., Thermochimica Acta **342** (1999) 67-72.
15. W. F. Zhang, Y. L. He et al., J. Phys. D: Appl. Phys. **33** (2000) 912-916.

Synthesis and XRD/PL Studies of Pure and Sb_2O_3 Doped ZnO Nanophases

N. Boulares[1], K. Guergouri[1], N. Tabet[2] and C. Monty[3]

[1]*Mentouri University, Constantine, Algeria*
[2]*King Fahd University of Petroleum and Minerals, Saudi Arabia*
[3]*CNRS/Promes, Odeillo 66120 Font-Romeu, France*
kaguergouri@yahoo.fr

Abstract. Pure and Sb_2O_3 (0 to 5% molar fraction) doped ZnO nanophases were synthesized using a sublimation-condensation method in a solar furnace. The initial and final powders were characterized by X-ray diffraction (XRD) and photoluminescence (PL) techniques. XRD results showed no significant change in the lattice parameters and the presence of a new phase $Zn_7O_2Sb_{12}$ in the highly doped micropowders but not in the nanopowders. The photoluminescence spectra showed a strong donor-acceptor pair (DAP) emission in the pure untreated ZnO micropowder which is drastically reduced in pure and doped nanopowders. The donor-bound excitonic band (DX) includes three well resolved peaks in the PL spectra of the doped micropowders while the spectra of doped nanopowders showed a broader band. Furthermore, the free exciton emission was absent in all doped samples.

Keywords: ZnO, Nanopowders, XRD, XPS, Photoluminescence.
PACS: 78.55, 78.67-n, 81.07

INTRODUCTION

Zinc oxide is still getting more and more importance because of its properties permitting it to be used in large domain of technological applications. The most important fields invested by ZnO are the chemistry, the environment, the electronics, the photovoltaic and the optoelectronics [1]. The research on ZnO is based now on the improvement of its properties used in each of these fields. In this work, we have synthesized pure and doped micro- and nanophases and investigated the effect of doping with Sb2O3 on the powders properties. For this purpose several techniques have been used including. X-Ray Diffraction (XRD), X photoelectron Spectroscopy (XPS) and photoluminescence (PL).

SYNTHESIS AND CHARACTERIZATION OF POWDERS

Johnson & Matthey powder of pure ZnO was used as starting material with micrometric grain size. Sb_2O_3 powder was used to prepare ZnO doped micropowders in the form pellets containing up to 5% Sb_2O_3 molar fraction. The mixtures were annealed at 1200 °C for 2 hours. The nanopowders were obtained by exposing the pellets under 660 Torr air pressure to a concentrated solar flux in a reactor placed at the focus of a parabolic mirror (Fig.1).

The sublimation-condensation occurred after 15 to 30 minutes and a fine powder was depositing on the cold walls of the glass balloon [2] and on a water-cooled porous stick (Fig.2). The XRD experiments were carried out using a copper anode (Cu Kα line). The analysis of the experimental data allowed us to determine the phases in the oxide mixtures, the lattice parameters and the average size of the crystallites.

CP929, *Nanotechnology and Its Applications, First Sharjah International Conference*
edited by Y. I. Salamin, N. M. Hamdan, H. Al-Awadhi, N. M. Jisrawi, and N. Tabet
© 2007 American Institute of Physics 978-0-7354-0439-7/07/\$23.00

The photoluminescence experiments were carried out at liquid helium temperature by means of a Fourier transform spectrometer (FTIR) BOMEM DA8 used in emission mode. Luminescence was detected by a silicon photodiode, a germanium photodetector cooled at liquid nitrogen temperature, or an InSb detector, according to the emission wavelength. Excitation light was provided by the 335.8 nm line of an argon-ion laser. The laser beam was focused on the crystal to a diameter of 200 μm.

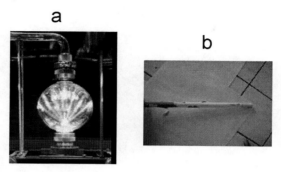

FIGURE 1. Glass balloon under focused solar light (a). Porous stick covered by ZnO nanopowder (b).

Figures 2 (a and b) and Figures 3 (a and b) show the diffraction patterns obtained from micropowders and nanopowders respectively for 2θ ranging from 30° to 40° and 45° to 65°.

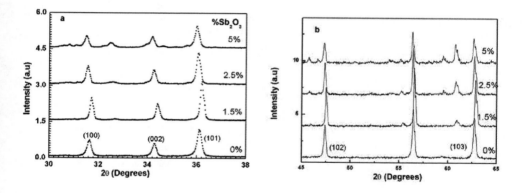

FIGURE 2. XRD patterns of pure and Sb_2O_3 doped micropowders.

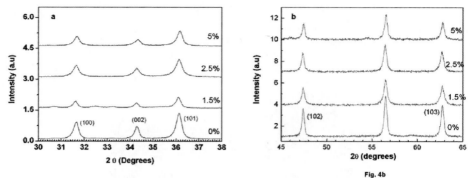

Fig. 4b

FIGURE 3. XRD patterns of pure and Sb_2O_3 doped nanopowders.

The main Bragg peaks in both patterns correspond to the known zincite structure of ZnO (würtzite). In the case of micropowders, low intensity extra peaks appear which can be attributed to the spinel phase $Zn_7O_2Sb_{12}$. It is worth noticing that this spinel phase does not appear in the diffraction patterns of nanopowders which suggests that nanophases could be considered as zincite type solid solution Zn1-xSbxO.

Figures 4a and 4b show the lattice parameters a and c determined from XRD for micro- and nanopowders. As we have not measured the actual concentrations of the micropowders, we have reported in Fig. 4 the variations of the parameters versus the Sb_2O_3 concentration in the initial mixtures. However, the c parameter remains quasi constant for all compositions for both micro and nanopowders (Fig.4b).

The determination of the average grain size is based on the Scherer formula applied by Langford in the analysis of the XRD peak width, taking into account the instrumental effect [3]. The Langford formula cannot be applied to micropowders because the effect of the grain size on the peak broadening is below the uncertainty. However, for the nanopowders, a slight increase of the average grain size as a function of the Sb concentration was observed.

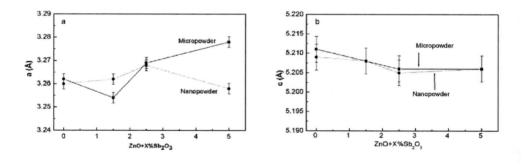

FIGURE 4. Variation of the lattice parameters a (Fig a) and c (Fig. b) versus the molar concentration of Sb_2O_3.

PHOTOLUMINESCENCE

The PL spectra are given in Fig. 5, 6 and 7 which show the excitonic region and a region of UV emission which are the regions of our interest.

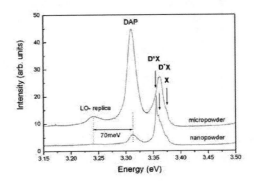

FIGURE 5. PL spectra of pure ZnO micropowder and nanopowder at T= 4.2 K.

The PL spectra obtained at Liquid Helium temperature from pure ZnO micropowder show the presence of five peaks (Fig. 5). The peak at 3.377 eV corresponds to free exciton transition (X). The two following peaks at 3.362 eV and 3.355eV are related to the donor-bound excitons (DX) [5]. The peak at 3.31 eV is associated with Donor-Acceptor Pair (DAP) emission. The 4th peaks at 3.241eV at 70 meV away from DAP line corresponds its longitudinal optical Phonon (LO) replica [6].

One can notice a significant reduction of the DAP peak in pure nanopowder. In addition, the D^+X component of the donor bound excitonic peak is reduced in the nanopowder. Figure 6 shows the PL spectra in the Sb-doped ZnO micropowders.

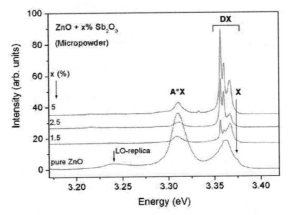

FIGURE 6. PL spectra of pure and doped ZnO micropowder at T= 4.2 K.

175

One can notice a significant reduction of the DAP peak in pure nanopowder. In addition, the D^+X component of the donor bound excitonic peak is reduced in the nanopowder. One can also observe that the peak of free exciton vanishes in the doped samples. The peaks related to donor bound excitonic band in the doped samples shows clearly three well resolved components of various intensities at 3.356eV, 3.360eV and 3.366eV and a significant reduction of the DAP emission. This reduction could be the result of the doping on the native donor concentration which is commonly identified as the doubly ionized oxygen vacancies.

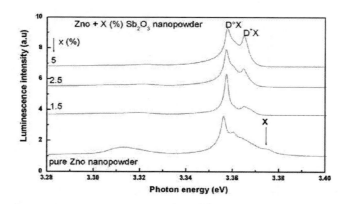

FIGURE 7. PL spectra of pure ZnO nanopowder and doped nanopowder at T= 4.2 K

However, we must notice that this decrease of the DAP emission could also be related to the annealing of the pellets at T=1200°C for two hours. In order to identify the actual cause of this reduction, the PL spectrum of a pure powder annealed in the same conditions is needed. Unfortunately, such spectrum was not recorded.

ACKNOWLEDGMENTS

The authors would like to thank F. Sibieude and A. Lusson for their assistance in XRD and PL experiments, respectively. N.T thanks King Fahd University of Petroleum and Minerals for its support.

REFERENCES

1. C. H. Kwong, H. K. Hong, and D. H. Yun, Sensors and Actuators B **24-25**, 610 (1995).
2. S. Major, S. Kumar, M. Bhatnagar, and K.L. Chopra, Appl. Phys. Lett. **49**, 394 (1986).
3. M. G. Ambia, M. N. Islam, and M. Obeidul Hakim, J. Mater. Sci. **27**, 5169 (1992).
4. N. Boulares, K. Guergouri, R. Zouaghi, N. Tabet, A. Lusson, F. Sibieude and C. Monty, Physica Stat. Solidi a. **201-10**, 2319 (2004).
5. Th. H. De Keijser, J. I. Langford, E. J. Mittmeijer and A. B. T. Vogels, J. Appl. Cryst **15**, 308 (1982).
6. B. R. Jackson and T. Pitman, U.S. Patent No. 6, 345 (2004) 224.
7. D. L. Davids, "Recovery Effects in Binary Aluminum Alloys" (Ph.D. Thesis, Harvard University, 1998).

A New Method of Preparation of Magnetite Nanoparticles from Iron Oxide Hydroxide

S. A. Kahani and M. Jafari

Department of Chemistry, University of Kashan, 87317-51167 Kashan, Iran

Abstract. There are only two basic ways to produce magnetite at low temperature: by partial oxidation of a Fe^{II} salt solution with oxidant under alkaline condition and by precipitation of a mixed Fe^{II}/Fe^{III} solution. At high temperature, the method involves reduction of iron oxide by a suitable reductant. Here we propose a new method of preparation of nanoparticle Fe_3O_4 from iron oxyhydroxides (goethite, akaganeite, lepidocrocite, feroxyhyte and ferrihydrite) or iron oxide (hematite) and ferrous salt in aqueous solution. Products characterized by X-ray powder diffraction, IR spectra and vibrating sample magentometery.

Keywords: Magnetite; Iron oxyhydroxide; Iron oxide.
PACS: 75.47.Lx; 75.75+a; 81.07-b

INTRODUCTION

Magnetite offers great potential for advancement in electronics, optioelectronics, magnetic storage, biomedical, ferrofluids, separation applications and magnetically guided drug carriers for targeting therapy [1, 2]. Recently, synthesis of magnetic materials on the nanoscale has become a topic in nanotechnology [3]. Only two basic ways to produce magnetite were reported in the literature [4]. At low temperature by partial oxidation of a Fe^{II} salt solution with oxidant under alkaline conditions at 90℃ and by precipitation of a mixed Fe^{II}/Fe^{III} solution with a Fe^{II}/Fe^{III} ratio of 0.5. The high temperature method of producing magnetite is by reducing hematite in a stream of H_2 at between 250-600 ℃ [2]. Here, a new method of preparation of magnetite is proposed. In this method magnetite is produced from a heterogeneous alkaline solution at 90℃ which contains iron oxyhyroxides or iron oxide and Fe^{II} with a suitable mole ratio. There are fifteen iron oxyhydroxide iron oxides and iron hydroxides known to date [2,4]. The new method is powerful in choosing a typical iron oxyhydroxide or oxides and results in different color, particle size and morphology of Fe_3O_4. Under the appropriate conditions, every iron oxide and iron oxyhydroxide can be converted into others. Here the conversion of goethite, akaganeite, lepidocrocite, feroxyhyte, ferrihydrite, and hematite into magnetite is important for developing new transformation routes. These interconversions are important because the magnetite has great potential in modern industry and nanotechnology.

EXPERIMENTAL

The Samples

Reagents used were of analytical grade and were obtained from Merck Co. All preparations were carried out under nitrogen and solvents were of technical grade, used after distillation. Iron oxyhyroxides (goethite, akaganeite, lepidocrocite, feroxyhyte and ferrihydrite) were prepared according to the procedure reported in [4,5]. Hematite was obtained from Merck Co. The Fe_3O_4 prepared here was dried at room temperature in air atmosphere to form a powder, to which we refer here by the powder samples (labeled henceforth by S_1 through S_6), were characterized by powder X-ray diffraction (XRD), infrared spectra (IR) and vibrating sample magentometery (VSM). XRD measurements were performed using a Philips Xpert pero MPD diffractometer with Cu Kα radiation in the range 2θ

CP929, *Nanotechnology and Its Applications, First Sharjah International Conference*
edited by Y. I. Salamin, N. M. Hamdan, H. Al-Awadhi, N. M. Jisrawi, and N. Tabet
© 2007 American Institute of Physics 978-0-7354-0439-7/07/$23.00

from 10 to 80, at room temperature. IR was obtained as KBr pellet in the range 4000 to 500 cm^{-1} using a Shimadzu FT-IR spectrometer. The vibrating sample magnetization (PAR-VSM155R) was used to evaluate the magnetic parameters.

Synthesis

(S$_1$): Magnetic from α-FeOOH (goethite)

A mass of 8.9 gram (0.1 mole) of goethite and 9.94 g (0.05 mole) FeCl$_2$ .4H$_2$O in 250 mL deionized water are placed in a 500-mL round-bottom flask with a side arm gas inlet tube (flushed with N$_2$) and a reflux condenser and a magnetic stirring bar is placed in the flask. Addition of 50 mL of 2 M NaOH is done with vigorous stirring. The reaction mixture is heated to the boiling point, and reflux is maintained for 1-2 hours. Precipitation of the Fe$_3$O$_4$ particles was formed and repeatedly washed with distilled water and filtered and dried at room temperature. Yield of 97%; X-ray diffraction (XRD) data in 2θ and relative intensity: 35.4235 (100); 30.0885 (29.06); 62.5028 (27.47); 43.0672 (18.53); 56.8920 (15.18); 18.2818 (7.15); 53.4050 (6.47) and 37.1051 (5.32). IR data (KBr) 575-600 cm^{-1} and 450 cm^{-1}. The magnetic values are saturation magnetization (M$_S$ = 72.20 emu/g), magnetic remanence (M$_R$ = 16.50 emu/g), coercive force (H$_C$ = 197.30 Oe).

(S$_2$): Magnetic from β-FeOOH (Akaganeite)

Preparation is the same as in (S$_1$) with 8.9 gram of β-FeOOH (Akaganeite) and 9.94 g FeCl$_2$ .4H$_2$O as reactants. Yield was 97%.

(S$_3$): Magnetic from γ-FeOOH (Lepidocrocite)

Preparation method was the same as in (S$_1$) with 8.9 gram of γ-FeOOH (Lepidocrocite) and 9.94 g FeCl$_2$ .4H$_2$O as reactants. Yield was 92%.

(S$_4$): Magnetic from δ-FeOOH (Feroxyhyte)

Preparation method was the same as in (S$_1$) with 8.9 grams of δ-FeOOH (Feroxyhyte) and 9.94 g FeCl$_2$ .4H$_2$O as reactants. Yield was 96.8%.

(S$_5$): Magnetic from Fe$_5$HO$_8$.4H$_2$O (Ferrihydrite)

Preparation method was the same as in (S$_1$) with 9.62 g (0.1 mole Fe^{3+}) Fe$_5$HO$_8$.4H$_2$O (Ferrihydrite) and 9.94 g (0.05 mole) FeCl$_2$ .4H$_2$O as reactants. Yield was 94.8%.

(S$_6$): from α-Fe$_2$O$_3$ (Hematite)

The preparation method same as (S$_1$) with 16 grams (0.1 mole) of α-Fe$_2$O$_3$ (Hematite) and 19.88 g (0.1 mole) FeCl$_2$ .4H$_2$O as reactants. Yield was 93.7%.

RESULTS AND DISCUSSION

Inter-conversions between the iron compounds are possible and often occur readily. An overview of the more frequent inter-conversion is reported in the literature [4]. Here we reported a new schematic presentation of transformation a pathway (fig. 1) from a different iron oxide and iron oxide hydroxide to magnetite. Since the advent of magnetic recording systems and other important fields of application, however, ferromagnetic iron oxides gained their importance quite rapidly. The new method is a feature in preparation of magnetite. Here the transformation of iron oxide to magnetite can be accompanied with the change in crystal system. Goetite and lepidocrocite have orthorhombic crystal structure. In inter-conversion to magnetite this changes to a cubic crystal system. Ferrihyrite and hematite have a trigonal crystal structure and also akaganiete and feroxyhyte, respectively, have monoclinic and hexagonal structures which all change to cubic. The patterns of iron oxides and oxihydroxides and products of the JCPDS database are included for comparison.

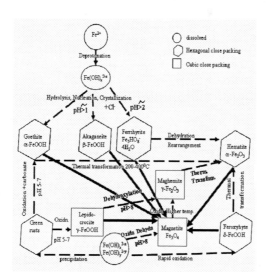

FIGURE 1. (a) Dashed line: schematic presentation of formation and transformation pathways in iron oxides reported, and (b) solid line: new pathways in magnetite formation.

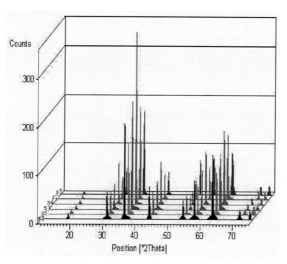

FIGURE 2. X-ray powder diffraction pattern of product samples (S_1- S_6).

X-ray Powder Diffraction

One of the best methods in characterization of these products is powder X-ray diffraction. A few years after the first successful application of X-ray diffraction to crystal structure determination, this technique was used to establish the major feature of the structure of magnetite [6]. The determination of the crystalline structure using X-ray diffraction is performed to obtain the crystalline lattice spacing of a sample. The Magnetite sample's X-ray diffraction pattern is shown in Fig. 2. The XRD pattern for the samples indicates that Fe_3O_4 is the dominant phase despite the broadening of the peaks, which is due to the size effect. X-ray diffraction method is applied not only to structure determination, but to the measurement of particle size. The crystal size can be calculated from the line broadening of the XRD pattern using the Scherrer formula shown in Eq. 1. The Equation uses the corrected reference peak width at angle θ, where λ is the X-ray wavelength, B is the corrected width of the XRD peak at half height and K is a shape factor which is approximated as 0.89 for magnetite [7]

TABLE 1. Particle diameter obtained from Scherrer's formula, for product samples (S_1 through S_6).

Sample	B (deg.)	θ (deg.)	D (nm)
S_1	0.0787	35.4235	123.35
S_2	0.23262	35.4388	40.99
S_3	0.1968	35.5564	49.28
S_4	0.2362	35.4968	41.00
S_5	0.1771	35.5532	54.77
S_6	0.0787	35.4343	123.30

$$D = \frac{0.89\lambda}{B\cos\theta} . \tag{1}$$

The mean crystallite size was calculated from XRD line broadening using the Scherrer relationship, in which D stands for the diameter of a particle of Fe_3O_4, using the data shown in Table 2.

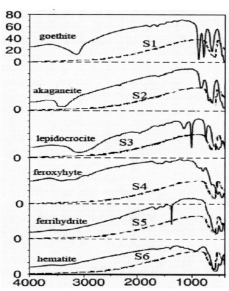

FIGURE 3. (a) Solid lines: IR spectra of reactants (goethite, akaganeite, lepidocrocite, feroxyhyte, ferrihydrite and hematite), and (b) Dash lines: IR spectra products (S_1- S_6).

Infrared Spectroscopy

There are various spectroscopic methods used to identify and characterize iron oxide. The most widely applied methods employ infrared techniques. For iron oxide, IR spectroscopy is useful as a means of identification. This technique also provides information about crystal morphology, and hence the nature of surface hydroxyl groups and adsorbed water. IR spectra of the various iron oxide and iron oxide hydroxide in comparison with magnetite samples are depicted in Fig. 3.

The IR spectrum of various magnetite sample show broad bands at 580 and 400 cm^{-1}. The IR of iron oxyhydroxides shows bands for surface OH groups, and bulk OH stretch and bending of OH groups. In iron oxyhydroxides converted to magnetite, absorption bands due to hydroxyl groups disappeared and the Fe-O broad band at 580 and 400 cm^{-1} appeared. IR spectrum of hematite showed bands at 662, 526 and 490 cm^{-1} for Fe-O stretch hematite and contains no structural OH groups.

Hysteresis

The most common way to represent the magnetic properties of a ferromagnetic material is a plot of magnetization (M) against field the strengths (H), the so-called hysteresis loop. The suitability of ferromagnetic materials for application is determined principally from characteristics shown by their hysteresis loops [8]. Such information includes: saturation magnetization (M_S), magnetic remanance (M_R), coercive force (H_C), of particle samples. The hysteresis loop data of particle samples S_1 through S_6, which was measured in the powder state at 300 ^{0}K is shown in Table 2.

The magnetization of bulk materials is inherently larger than for nanoparticles of the respective materials. Experimental values for the saturation magnetization of the samples are in the range 46.10 to 74.73 emu/g, which is significantly less than that of the bulk magnetization (M_S for the bulk material is 92 emu/g [9]). The value of M_S itself will be regarded simply as a constant of the material. Several factors affect the shape of the hystersis loop and the M_S value. One factor which may strongly affect is magnetic anisotropy. A plot of the saturation magnetization of samples observed in Table 2 versus particle size obtained by powder diffraction (given in Table 1), exhibits a linear correlation magnetic saturation and particle size (see Fig. 4).

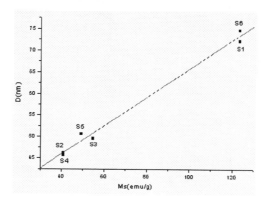

FFIGURE 4. Linear dependence of the saturation magnetization of product samples (S_1- S_6) and nanoparticle size.

Several researchers using a variety of techniques have investigated the causes for the observed reduction in magnetization in fine magnetic particles. Results suggest that particle surface defects and order-disorder structural characteristics both influence the magnetic properties [10]. The coercivity is used to distinguish between hard and soft magnetic materials. The remanence is dependent on the saturation magnetization and therefore in permanent magnet materials High M_S, H_C and H_R is necessary. There are various magnetic methods available for information storage. The most important of these are Magnetic tape recording, which is widely used for both audio and video recording, and magnetic disk recording, come in two categories: floppy disks and hard disks. The magnetic recording media must have high saturation magnetization to give as large a signal as possible during the reading process. The coercivity must be sufficient to prevent erasure, but small enough to allow the material to be reused for

181

recording. Here the hysteresis magnetization of samples S_1 - S_6 shows suitable values for the M_S, M_R and H_C parameters and can be used to good effect in magnetic recording.

TABLE 2. Magnetic properties of product samples.

Sample	M_S (emu/g)	M_R (emu/g)	H_C (Oe)
S_1	72.20	16.50	197.30
S_2	46.10	17.00	164.40
S_3	50.66	15.90	130.68
S_4	45.70	14.50	177.60
S_5	49.60	17.94	183.70
S_6	74.73	25.28	223

CONCLUSIONS

In this article we presented the new method of preparation of Fe_3O_4 nanoparticles from various iron oxides and iron oxyhydroxide. XRD and magnetization analyses have shown that the Fe_3O_4 particles prepared are different in particle size. We also carefully studied the hysteresis loop and the dependence of the magnetic properties on reactants and preparation method. With change in preparation of the raw materials, the saturation magnetization M_S is increased due to the increase in particle size. The IR spectra and powder X-ray diffraction pattern of samples have shown that all samples had a single-phase magnetite structure

REFERENCES

1. C. Bergemann, D. Müller-Schulte, J. Oster, L. à Brassard, and A.S. Lübbe, J. Magn. Magn. Mater. **194** (1999) 45.
2. R. M. Cornell, U. Schwertmann, the Iron Oxides (Wiley, Weinheim, 2003).
3. P. Tartaj, M. P. Morales, S. V. Verdaguer, T.G. Carrenon and C. J.Serna, J. Phys. D: Apply. Phys. **36** (2003) R 182.
4. U. Schwertmann and R. M. Cornell, Iron Oxides in the Laboratory (VCH, 2000).
5. B. Weckler and H.D. Lutz, Eur. J. Solid State Inorg. Chem. **35** (1998) 531.
6. B. D. Cullity, Element of X-ray Diffraction (Addison Wesley, Reading, MA, 1974).
7. H. P. Klug and L. E. Alexander, X-ray diffraction procedures for polycrystalline and amorphous materials (J. Wiley and Sons, New York, 1974) p. 966.
8. D. Jiles, Introduction to Magnetism and Magnetic Materials (Chapman and Hall, 1991) p. 71.
9. Z. L. Liu, Y. J. Liu, K. L. Yao, Z. H. Ding, J. Tao, and X. Wang, Journal of Materials Synthesis and Processing **10** (2002) 83.
10. M. P. Morales, M. Andres-Verges, S. Veintemillas-Verdaguer, M. I. Montero, C. J. Serna, J. Magn. Magn. Mater. **203** (1999) 146.

Deposition of Magnetite Nanoparticles in Activated Carbons and Preparation of Magnetic Activated Carbons

S. A. Kahani, M. Hamadanian and O. Vandadi

Department of Chemistry, University of Kashan, 87317-51167 Kashan, Iran

Abstract. Magnetic activated carbons (MACs) for gold recovery from alkaline cyanide solutions have been developed by mixing a magnetic precursor with a carbon source, and treating the mixture under controlled conditions. As would be expected, these activated carbons have high specific surface areas due to their microporous structure. In addition, the small particle size of the MACs produced allows rapid adsorption of gold in solution, and the magnetic character of these MACs enables recovery from suspension by magnetic separation.

Keywords: Magnetite; Magnetic activated carbons; gold recovery.
PACS: 75.47.Lx; 75.75+a; 81.07-b

INTRODUCTION

Magnetic separation has been developed as a recovery and pollution-control process for many environmental and industrial problems. Application of efficient magnetic filtration to petroleum-related decontamination and waste treatment operations is attractive because it can provide rapid removal of contaminants from aqueous waste streams [1]. The simplest method was to mix an extremely large excess of magnetite oil (fatty acid) by weight so that the oil absorbed onto the surface of the magnetic powder. Activated carbon has a long-standing history as an absorbent and extraction agent of Precious metals such as gold from its ore [2-3]. Thus, conversion of activated carbon into a magnetically-active material would provide an excellent magnetic filtration aid. Here a method proposed for the preparation of porous carbon that was loaded with magnetite particles. Magnetic activated carbons (MACs) for gold recovery from alkaline cyanide solutions have been developed by mixing a magnetic precursor with a carbon source, and treating the mixture under controlled conditions. As would be expected, these activated carbons have high specific surface areas due to their microporous structure. In addition, the small particle size of the MACs produced allows rapid adsorption of gold in solution, and the magnetic character of these MACs enables recovery from suspension by magnetic separation. Magnetic carbon active applied as magnetic carriers that could be used in the recovery of gold from cyanide solutions. Based on their results, in this investigation the extraction of gold up to 99% can be achieved.

EXPERIMENTAL

Reagents used were of analytical grade and were obtained from Merck Co. All the preparations were carried out under the nitrogen and solvents were of technical grade, used after distillation. XRD measurements were performed using a Philips X,pert pero MPD diffractometer with Cu Kα radiation in the range 2θ from 10 to 80, at room temperature. IR was obtained as KBr pellet in the range 4000 to 500 cm^{-1} using Shimadzu FT-IR spectrometer. The vibrating sample magnetization (PAR-VSM155R) was used to evaluate the magnetic parameters. Scanning electron micrographs (SEM) were recorded, without sample coating, by JEOL JSM-5600LV.

CP929, *Nanotechnology and Its Applications, First Sharjah International Conference*
edited by Y. I. Salamin, N. M. Hamdan, H. Al-Awadhi, N. M. Jisrawi, and N. Tabet
© 2007 American Institute of Physics 978-0-7354-0439-7/07/$23.00

Preparation of Magnetite Using a Homogenous Method

The solutions containing 2.72 g of $FeCl_3$ and 1 g of $FeCl_2$ were mixed at a certain molar ratio. Then 20 mL of NH_4OH 25% was slowly injected into the mixture of $FeCl_3$ and $FeCl_2$ under vigorous stirring at temperature 60-70 ^0C. Before the reaction, N_2 gas was injected through the reaction medium to prevent possible oxidation. After precipitation the Fe_3O_4 particles were repeatedly washed and filtered before drying at room temperature [4]. The chemical reaction of Fe_3O_4 precipitation can be described as follows

$$Fe^{2+} + 2Fe^{3+} + 8OH^- \rightarrow Fe_3O_4 + 4H_2O. \tag{1}$$

Doping of Magnetite in Activated Carbon

A solution containing 2 g of magnetite and 2 g of commercial activated carbon was prepared in 40 ml distilled water under vigorous stirring at room temperature. After 5 hours the doping is complete. Filtering and washing and drying in an oven at 100 ^0C are carried out.

Preparation of Magnetite Using a Heterogeneous Method

A mass of 8.9 gram (0.1 mole) of goethite, 2 g activated carbon and 9.94 g (0.05 mole) $FeCl_2 .4H_2O$ in 250 mL deionized water are placed in a 500-mL round-bottom flask with a side arm gas inlet tube (flushed with N_2) and a reflux condenser and a magnetic stirring bar are placed in the flask. This is followed by adding 50 mL of 2 M NaOH with vigorous stirring. The reaction mixture is heated to the boiling point, and reflux is maintained for 1-2 hours. The reaction should be carried out under N_2. During the transformation the pH falls to 8-9 and a black precipitate is formed in activated carbon. After the reaction is complete the product was repeatedly washed with distilled water and filtered and dried in an oven at the temperature 100 ^0C. The chemical reaction of Fe_3O_4 precipitation can be described as follows

$$Fe^{2+} + 2FeOOH + 2OH^- \rightarrow Fe_3O_4 + 2H_2O. \tag{2}$$

Gold Adsorption on MACs

A 1-liter solution containing 10 mg of gold as the aurodicyanide complex $Au(CN)_2$ in 1 liter of solution (10 ppm Au) at pH 11, a free sodium cyanide concentration of 0.02 M, and at room temperature (25°C) was stirred at 1500 RPM with an overhead motor. One gram of magnetic activated carbon was added to the system. Samples of 10 mL were taken at specific times with a syringe, the solution was filtered, and the amount of gold in solution was determined by AAS. Based on the amount of gold in solution after specific adsorption times, the % of gold adsorption at each time could be calculated.

RESULTS AND DISCUSSION

Magnetic particles have shown great promise as a base material for engineering carriers in separation science and technology. Magnetic particles used in these areas are called magnetic carriers [5]. The objective of magnetic carrier technology is to confer the magnetic property to a naturally nonmagnetic target so that the target can be separated from the stream using magnetic separators. This is particularly important for separations in a complex multiphase system. To make a target magnetic, the target has to be recognized by and attached to a magnetic carrier. This requires the magnetic particles to be usually on a surface or produced in situ formation of a magnetic core in a structured template such as active carbon, porous silica, or polymer. The formation of magnetic cores in porous media is important in magnetic separation. Examples have been given for magnetic carriers of tailored functionalities in a variety of applications including industrial effluent detoxification, metal recovery, biological cell separation, drug delivery, and the preparation of biological sensors. Activated carbon was introduced as an adsorbent for the recovery of precious metals such as gold and silver from cyanide solutions. In recent years activated carbon recovery systems have gained very wide acceptance in the gold industry. Here we proposed the doping of magnetic particles in activated carbon. Magnetic activated carbon applied as a magnetic carrier that could be used in the recovery of gold from cyanide solutions. Activated carbon is a highly porous material with very large intra-particulate area per unit of mass. This requires the magnetic particles for preparation of Magnetic activated carbon to be extremely small, in the nanometer range. The magnetite particles and Magnetic activated carbon prepared are characterized by Fourier transform infrared spectroscopy (FTIR), X-ray powder diffraction (XRD), vibrating sample magnetometry (VSM) and scanning electron microscopy (SEM) and X-Ray Microanalysis.

Infrared Spectroscopy

The most widely applied techniques for characterizing of iron oxide involve infrared. For iron oxide, IR spectroscopy is useful as a means of identification. This technique also provides information about crystal morphology, and hence the nature of surface hydroxyl groups and adsorbed water. The IR spectra of various magnetite samples show broad bands at 580 and 400 cm^{-1}. The IR spectrum of magnetic activated carbon (MAC$_1$) shows broad bands at 1185, 598 and 401 cm^{-1}. The intense band in the region 1000-1185 cm^{-1} can be attributed to stretching of C-C and C-O in activated carbon and 581 and 401 cm^{-1} can be attributed to deposited magnetite. Magnetic activated carbon (MAC$_1$) shows broad bands in the region 1000-11200, 614 and 420 cm^{-1}. The intense bands at 1000-1200 cm^{-1} can be attributed to stretching of C-C and C-O in activated carbon and 614 and 420 cm^{-1} can be attributed to F-O stretching deposited magnetite. Infrared spectroscopy demonstrated the presence of magnetite in activated carbon phase.

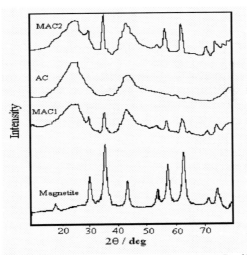

FIGURE 1. X-ray diffraction pattern of magnetite, MAC$_1$, AC and MAC$_2$.

X-ray Powder Diffraction

One of the best methods in characterization of these products is powder X-ray diffraction. A few years after the first successful application of X-ray diffraction to crystal structure determination, this technique was used to establish the major feature of the structure of magnetite [6]. The determination of the crystalline structure using X-ray diffraction is performed to obtain the crystalline lattice spacing of a sample. The crystal structure contains symmetry elements along planes, axes or centers. The intercept with any plane of a symmetry axis is defined by the reciprocal values of the intercepts, hkl, known as Miller indices. Magnetite and Magnetic activated carbon samples X-ray diffraction pattern are shown in Fig. 1.

X-ray powder diffraction showed that the resulting material contained nanocrystalline magnetite (see Figure 1) XRD patterns for the samples indicate that Fe_3O_4 is deposited in activated carbon. X- ray diffraction method is applied not only to structure determination, but to the measurement of particle size. The crystal size can be calculated from line broadening of the XRD pattern using the Scherrer formula shown in Equation 3. The Equation uses the corrected reference peak width at angle θ, where λ is the X-ray wavelength, B is the corrected width of the XRD peak at half height and K is a shape factor which is approximated by 0.89 for magnetite [7]

$$D = \frac{0.89\lambda}{B\cos\theta}.$$

(3)

The mean crystallite size was calculated from XRD line broadening using the Scherrer relationship. The diameter of particle of Fe_3O_4 are shown in Table 1.

TABLE 1. Particle diameter obtained from Scherrer's formula, for product samples M_1 and M_2.

Sample	B (deg.)	θ (deg.)	D (nm)
M_1	0.5314	35.5306	18.26
M_2	0.0787	35.4235	123.35

Hysteresis

The most common way to represent the magnetic properties of a ferromagnetic material is plots of magnetization (M) against the field strengths (H), the so-called hysteresis loop. The suitability of ferromagnetic materials for application is determined principally from characteristics shown by their hysteresis loops [8]. The magnetization properties were investigated using vibrating sample magnetometery (VSM) which provided fundamental magnetic behavior quantification. One important parameter of the magnetization properties is saturation magnetization (M_S) of the particle samples. The hysteresis loop data of samples M_1 and M_2 particle, which was measured in the powder state at 300 ^{0}K is shown in Table2.

TABLE 2. Magnetic properties of product samples.

Sample	M_S (emu/g)	% Magnetite
M_1	50.27	-------
M_2	72.20	-------
MAC_1	28.33	13.20
MAC_2	15.45	11.80

The magnetization of bulk materials is inherently larger than for nanoparticles of the respective materials. Experimental values for the saturation magnetization of our samples are 50.27 emu/g for M_1 and 72.20 emu/g for M_2, significantly less than that of bulk magnetization, which is 92 emu/g [9]. The value of M_S itself will be regarded simply as a constant of the material. Several factors affect the shape of the hystersis loop and M_S value. One factor which may strongly affect is particle size. Several researchers using a variety of techniques have investigated the causes for the observed reduction in magnetization in fine magnetic particles. Results suggest that particle surface defects and order-disorder structural characteristics both influence the magnetic properties. The values for the saturation magnetization of our samples are 28.33 emu/g for MAC_1 and 15.45 emu/g for MAC_2 (see Fig. 2). These are significantly less than the magnetization of the magnetic samples. This is due to strong dependence on the particle size, percent of magnetite in sample, particle surface defects and order-disorder structural dependence.

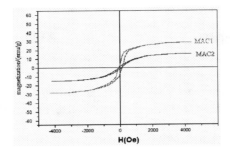

FIGURE 2. Hysteresis loop of magnetic activated carbon samples, MAC_1 and MAC_2.

Scanning Electron Microscopy (SEM)

The morphology and size distribution of particles are most easily obtained by SEM. SEMs showed the development of porosity of the magnetic activated carbon, and nano-crystals of inorganic magnetite deposits are clear. Figure 3 shows the SEM photographs of magnetite in activated carbon samples MAC_1 and MAC_2, respectively.

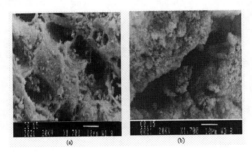

FIGURE 3. SEM photographs of magnetic activated carbon (a) MAC_1, and (b) MAC_2.

On the other hand, microanalysis is the determination of the composition of the specimen on a microscopic scale. In an electron microscope, this is done by irradiating with a beam of electrons and analyzing the radiation emitted. In Energy Dispersive X-Ray (EDX) qualitative identification of the elements present in a sample and their spatial distribution are possible. The minimum detection limits vary around about 0.1 percent by weight. The EDX spectra of activated carbon and magnetic activated carbon (MAC_1) are show in Fig. 4.

The results of X-ray diffraction analysis according to peak height show a strong $K\alpha$ peak for carbon and a weak $K\alpha$ peak for traces of oxygen in activated carbon. In magnetic activated carbon, oxygen $K\alpha$ peak increases and also Fe-Lα, Fe-Kα and Fe-Kβ peaks appear for magnetite deposition.

Results show that gold adsorption increases in MAC_2. The percent adsorption of gold on activated carbon and homogenous magnetic activated carbon (MAC_1) and heterogeneous magnetic activated carbon (MAC_2) are 100%, 77%, and 95%, respectively.

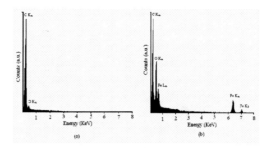

FIGURE 4. The EDX spectra of (a) activated carbon, and (b) magnetic activated carbon.

The formation of micropores and the extent of specific surface area are considered to be important factors for adsorption. According to literature, the pores of activated carbon are classified into three groups: micropores (dp<2nm), mesopores (2nm<dp<50nm) and macropores (dp>50nm), where dp is the pore width [10]. Activation widens the micropores resulting in more developed porosity. During adsorption, macro and mesopores allow rapid transport of adsorbate into the interior of carbon, for subsequent diffusion into the micropore volume. Consequently, a well developed porous network in all pore size ranges results in improved adsorption properties of the product. The porosity is dependent on the magnetic particle size on the deposit of activated carbon. In MAC_1 magnetite nanoparticle can occupy mesopores and in MAC_2 only deposit in macropores is possible. A better way to produce magnetic activated carbon is a heterogeneous mixture of activated carbon and the magnetic precursor.

REFERENCES

1. J. Svoboda J., *Magnetic Methods for Treatment of Minerals* (Elsevier, 1987).
2. E. Bernardo, R. Egashira, J. Kawasaki, Carbon **35** (1997) 1217.
3. R. C. Bansal, D. B. Donnet and F. Stoekli, Active Carbon (Marcel Dekker, New York, 1988).
4. Z. L. Liu, Y. J. Liu, K. L. Yao, Z. H. Ding, J. Tao, and X. Wang, Journal of Materials Synthesis and Processing **10** (2002) 83.
5. G. Moffat, R. A. Williams, C. Webb and R. Stirling, Miner. Eng. **7** (1994) 1039.
6. B. D. Cullity, *Element of X-ray Diffraction* (Addison Wesley, Reading, MA 1974).
7. H. P. Klug and L. E. Alexander, X-ray Diffraction Procedures for Polycrystalline and Amorphous Materials (J. Wiley and Sons, New York, 1974) p. 966.
8. D. Jiles, *Introduction to Magnetism and Magnetic Materials* (Chapman and Hall, 1991) p. 71.
9. M. P. Morales, M. Andres-Verges, S. Veintemillas-Verdaguer, M. I. Montero, C. J. Serna, J. Magn. Magn. Mater. **203** (1999) 146.
10. P. Galiatsatou, M. Metaxas and V. K. Rigopoulou, Mikrochimica Acta **136** (2001) 147.

BIO-NANOTECHNOLOGY

Nanobiomaterials for Electrochemical Biosensors

M. Pumera

ICYS, National Institute for Materials Science, 1-1 Namiki, Tsukuba, Ibaraki, Japan
PUMERA.Martin@nims.go.jp
http://www.nims.go.jp/icys/02members/pumera

Abstract. I will discuss main techniques and methods which use nanoscale materials for construction of electrochemical biosensors with emphasis on methods developed by myself and my coworkers. Described approaches include carbon nanotube based electrodes relying on double wall and multiwall carbon nanotubes, novel binding materials and mass production technology; and nanoscale materials as biomolecule tracers, including gold nanoparticles for DNA detection. Specific issues related to electrochemistry of nanoscale materials will be discussed. Various applications for genomic and proteomic analysis will be described.

Keywords: carbon nanotubes, double wall carbon nanotube, composites, nanoparticles, electrochemistry, biosensor, sensor.
PACS: 82.45.Tv, 87.15.Tt, 81.07.–b, 73.63.Fg

CARBON NANOTUBE ELECTRODES

The advantages of carbon nanotubes, such as high surface area, favorable electronic properties and electrocatalytic effect attracted very recently considerable attention for the construction of electrochemical enzyme biosensors [1,2]. It is known that when multiwall carbon nanotubes (MWCNT) are oxidized by nitric acid, it introduces defects to their outer graphene sheet which results to improved heterogeneous electron transfer while inner graphene layers acts as electrical wire. An important question then arises as to what minimum number of graphene layers would allow such a function. We tested the electrochemical properties of the smallest multiwall carbon nanotube – the double wall carbon nanotube (DWCNT) and compared them to their single wall (SWCNT) counterparts [3], see Figure 1. The double and single wall carbon nanotube materials were characterized by Raman spectroscopy, scanning and transmission electron microscopy and electrochemistry. The electrochemical behavior of DWCNT film electrodes was characterized by using cyclic voltammetry of ferricyanide and NADH. It was shown that while both DWCNT and SWCNT were significantly functionalized with oxygen containing groups, double wall carbon nanotube film electrodes show a fast electron transfer and substantial decrease of overpotential of NADH when compared to the same way treated single wall carbon nanotubes. The results presented above reveals that outer wall of oxidized double wall carbon nanotube can provide active sites for effective oxidation of biomolecules while inner wall of nanotube acts as 1D nanowire. This property is similar to the one found in the multiwall carbon nanotubes, however, double wall nanotubes provide much smaller diameter [3]. We further showed the use of DWCNT as platform for non-covalent adsorption of redox protein glucose oxidase for construction of binder-less glucose biosensor [4].

CP929, Nanotechnology and Its Applications, First Sharjah International Conference
edited by Y. I. Salamin, N. M. Hamdan, H. Al-Awadhi, N. M. Jisrawi, and N. Tabet
© 2007 American Institute of Physics 978-0-7354-0439-7/07/$23.00

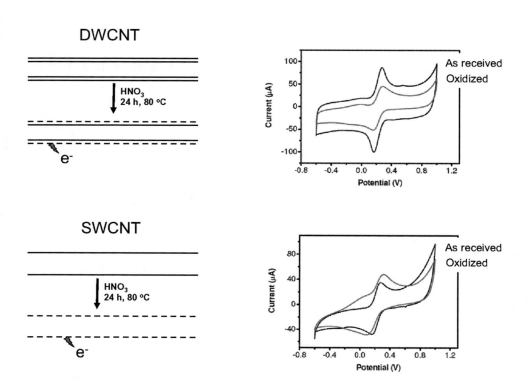

FIGURE 1. Schematic drawing of DWCNT and SWCNT graphene sheets before and after oxidation with concentrated nitric acid (left). Corresponding cyclic voltammetry of 5 mM potassium ferricyanide at as received and oxidized nanotubes. Reprinted with permission from [3].

Next we investigated employing carbon nanotubes for construction of rigid biosensors. We constructed glucose rigid carbon nanotube biosensors by incorporating carbon nanotubes and glucose oxidase in epoxy matrix [5, 6]. Further more, we persuaded the mass production route and we constructed screen-printed CNT sensors, based on thick-film fabrication, which are mechanically stable with good resistance to mechanical abrasion and they offer possibility of large scale mass production of highly reproducible low-cost electrochemical biosensors [7]. CNT matrix allows easy incorporation of enzyme in screen-printed electrode, as we demonstrated recently on example of horseradish peroxidase in connection to MWCNT and polysulfone binder [7] (see Figure 2). The apparent Michaelis-Menten constant K_M^{app} was calculated to be 0.71 mM. This K_M^{app} indicates that the enzyme immobilized in the carbon nanotube/polysulfone biocomposite keeps its activity with a very low diffusion barrier [7].

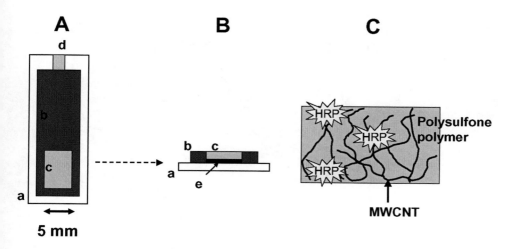

FIGURE 2. The MWCNT/polysulfone screen-printed thick-film electrochemical detector, top view; (B) Cross section of the detection area of MWCNT/polysulfone screen-printed detector; (C) schematic drawing of showing structure of HRP/MWCNT/PS composite; a) polycarbonate substrate, b) insulator layer, c) MWCNT/polysulfone conducting composite, d) silver contact for the working electrode, e) carbon ink contact layer. Reproduced with permission of The Royal Society of Chemistry from [7].

GOLD NANOPARTICLES AS ELECTROCHEMICAL TAGS

We described a novel gold nanoparticle-based protocol for detection of DNA hybridization based on a magnetically trigged direct electrochemical detection of gold quantum dot tracers [8]. It relies on binding target DNA (here called DNA1) with Au_{67} quantum dot in a ratio 1:1, followed by a genomagnetic hybridization assay between Au_{67}-DNA1 and complementary probe DNA (here called DNA2) marked paramagnetic beads. Differential pulse voltammetry was used for a direct voltammetric detection of resulting Au_{67} quantum dot-DNA1/DNA2-paramagnetic bead conjugate on magnetic graphite-epoxy composite electrode (Figure 3). The characterization, optimization, and advantages of the direct electrochemical detection assay for target DNA were demonstrated. The two main highlights of presented assay are (1) the direct voltammetric detection of metal quantum dots obviates their chemical dissolution and (2) the Au_{67} quantum dot-DNA1/DNA2-paramagnetic bead conjugate does not create the interconnected three-dimensional network of Au_{67}-DNA duplex-paramagnetic beads as previously developed nanoparticle DNA assays, pushing down the achievable detection limits [8].

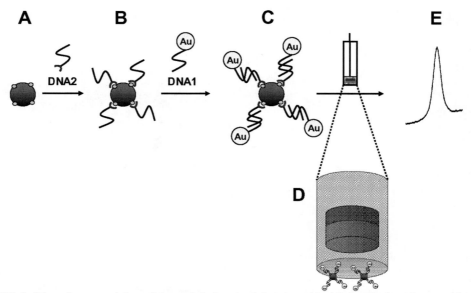

FIGURE 3. Schematic representation of the analytical protocol (not in scale): (A) introduction of streptavidin coated paramagnetic beads; (B) immobilization of the biotinylated probe (DNA2) onto the paramagnetic beads; (C) addition of the *1:1* Au_{67}-DNA1 target; (D) accumulation of Au_{67}-DNA1 / DNA2-paramagnetic bead conjugate on the surface of magnetic electrode; (E) magnetically trigged direct DPV electrochemical detection of gold quantum dot tag in Au_{67}-DNA1 / DNA2-paramagnetic bead conjugate. Reprinted with permission from [8].

ACKNOWLEDGMENTS

M. P. is grateful to the Japanese Ministry for Education, Culture, Sports, Science and Technology (MEXT) for funding through ICYS programme.

REFERENCES

1. M. Pumera, S. Sánchez, I. Ichinose, J. Tang, *Sensors Actuators B*, **123**, 1195-1205 (2007)
2. A. Merkoçi, M. Pumera, X. Llopis, B. Pérez, M. Valle, S. Alegret, *Trends Anal. Chem.* **24** 826-838 (2005).
3. M. Pumera, *Nanoscale Res. Lett.* **2**, 87-93 (2007).
4. M. Pumera, B. Smid, *J. Nanosci. Nanotech.* 2007, in press, doi:10.1166/jnn.2007.846.
5. B. Peréz, M. Pumera, M. Valle, A. Merkoci, S. Alegret, *J. Nanotech. Nanosci.* **5** 1694–1698 (2005).
6. M. Pumera, A. Merkoci, S. Alegret, *Sensors Actuators B* **113** 617–622 (2006).
7. S. Sánchez, M. Pumera, M., E. Cabruja, E., Fàbregas, *Analyst*, **132**, 142-147 (2007).
8. M. Pumera, M. T. Castaneda, M. I. Pividori, R. Eritja, A. Merkoci, S. Alegret, *Langmuir* **21**, 9625-9629 (2005).

Anti-Bacterial Self-Assembled Nanotubes of Cyclic D, L-α-Peptides

M. Al-Sayah[a] and M. R. Ghadiri[b]

[a]*Department of Biology and Chemistry, American University of Sharjah, P.O. Box: 26666, Sharjah, UAE*
[b]*Departments of Chemistry and Molecular Biology and the Skaggs Institute for Chemical Biology, The Scripps Research Institute, La Jolla, California, 92037, USA*
malsayah@aus.edu

Abstract. Cyclic D, L-α-peptides can be designed to spontaneously partition into lipid and cell membranes where they self-assemble into membrane-permeating nanotubes and result in disruption of membrane potentials leading to rapid cell death. The self-assembly of the flat ring-shaped cyclic D, L-α-peptides in lipid membranes is based on the formation of intermolecular hydrogen-bonds through the perpendicular backbone amide groups. Single channel conductance measurements and proton-transfer assays have been demonstrated the ability of such constructs to act as trans-membrane channels while dye-release assays have been employed to determine the pore size created by the self-assembled structures. The dynamics of the self-assembly of pyrene-labeled cyclic peptides into nanotubes in lipid membranes were monitored through the change in the fluorescence of pyrenes and the bio-activity is studied through MIC assays.

Keywords: Cyclic peptides, nanotubes, antibacterial, self-asssembly.
PACS: 87.83.+a

INTRODUCTION

Cyclic peptides made up of an even number of alternating D- and L-α-amino acid residues adopt flat ring-shaped conformations in which all amide backbone functionalities reside approximately perpendicular to the plane of the ring structure. In this conformation, the spatial disposition and juxtaposition of amide hydrogen-bond donor and acceptor sites on either side of the ring structure allows the peptide subunits to stack under conditions that favor hydrogen bonding to form uniformly-shaped and contiguously hydrogen-bonded β-sheet-like nanotube ensembles (Figure 1). [1-14] All peptide side chains lie on the outside of the tube due to steric considerations and the alternating amino acid configuration. By appropriate choice of amino acid sequence, cyclic D, L-α-peptides can be designed to spontaneously partition into lipid membranes and self-assemble into functional pores that enable ion and small molecule transport across the membrane. [2,3,14].

CHRACTERIZAION OF NANOTUBES

Solid-State Peptide Nanotubes

The internal diameter of cyclic D, L-α-peptide nanotubes can be controlled by simply varying the size of the peptide ring. The aggregates formed by octapeptide cyclo[-(l-Gln-d-Ala-l-Glu-d-Ala)$_2$-] has a van der Waal internal diameter of 7 Å while that of cyclo[-(l-Gln-d-Ala-l-Glu-d-Ala)$_3$-] is 13 Å. [14, 15] The microcrystalline aggregates formed by

CP929, *Nanotechnology and Its Applications, First Sharjah International Conference*
edited by Y. I. Salamin, N. M. Hamdan, H. Al-Awadhi, N. M. Jisrawi, and N. Tabet
© 2007 American Institute of Physics 978-0-7354-0439-7/07/$23.00

controlled acidification of the former were thoroughly examined by electron microscopy, electron diffraction, FT-IR spectroscopy, and crystal structure modeling. These analyses strongly support the formation of the expected structure in which the ring-shaped subunits stack to form hollow tubes. The unit cell parameters of solid-state nanotubular assemblies, obtained from cryoelectron microscopy and electron diffraction analyses, are in full agreement with the expected tubular structures of cyclic octapeptides. [6] Finally, the high stability of the aggregates formed from the cyclic D, L-α-peptide subunits and their low solubility suggests significant cooperative self-assembly process of the monomeric units[14].

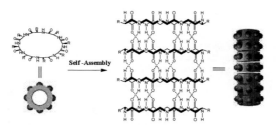

FIGURE 1. Schematic representation of the self-assembly of cyclic D, L-□-peptides. (Side chains were removed for clarity)

Self-Assembling Peptide Nanochannels

Cyclic D, L-α-peptides bearing appropriate hydrophobic side chains partition into nonpolar lipid bilayers and self-assemble to form nanochannels. Single-channel conductance measurements, glucose transport studies and fluorescence proton-transport assays have demonstrated the ability of such constructs to act as transmembrane channels.[2,3,14] The self-assembling ion channels of cyclic octapeptides display transport activities also for H^+, K^+ and Na^+.

The analysis of spectroscopic results of polarized attenuated total reflectance (ATR) and transmission Fourier transform infrared (FT-IR) of the complexes formed from multiple lipid bilayers and peptides supports this model of the peptide nanotubes as the active channel species.[9] The pore size created by the self-assembled structures has also been studied using dye-release assays.[13] All these studies have indicated that the mode of membrane permeation depends upon the amino acid sequence comprising the cyclic D, L-α-peptide units.

Fluorescence Studies of the Self-Assembly of Pyrene-Peptides

Pyrene units are utilized as the spectroscopic handle to monitor the self-assembly of cyclic D, L-α-peptide rings. Monomer pyrenes exhibit characteristic fluorescence emission (λ_{em} = 397 nm) upon excitation with UV light (λ_{ex} = 340 nm). However, cofacial interaction between one pyrene ring in its excited state with a second ring in its ground state at a 3.5 Å distance leads to quenching of the monomer fluorescence and the formation of excimers (λ_{em} = 470 nm). [16] Since the inter-unit separation distance of cyclic D,L-α-peptides assembled via β-sheet interactions is 4.8 Å, the assembly of pyrene-modified cyclic peptides places the pyrene rings in close proximity and enhances the formation of the excimer (Figure 2).

Solutions of pyrene-labeled cyclic peptides exhibit fluorescence spectra characteristic of monomer pyrene. Upon the introduction of a lipid bilayer, initial enhancement in the pyrene monomer fluorescence is observed due to the partition of the peptides into the hydrophobic membrane. As time progresses, a decrease in the monomer emission at 397 nm concomitant with an increase of the excimer emissions at 470 nm is noticed (Figure 3). This increase in the excimer emission provides evidence that the peptides self-assemble slowly upon partitioning into the membrane. The rate and the extent of the self-assembly of these peptides increase as the peptide-to-lipid ratio increases. This suggests that the elf-assembly of the pyrene-labeled cyclic peptides is a corporative process in which the partition and the assembly are oncurrent processes at high peptide-to-lipid ratios.

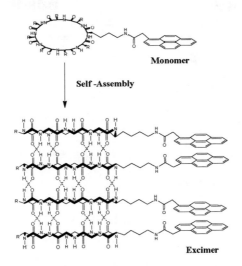

FIGURE 2. The self-assembly of pyrene-labeled cyclic D, L-α-peptides.

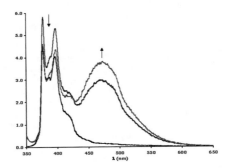

FIGURE 3. Fluorescence spectra of c[QXHRWLWK] (5µM) before (black) and after the addition of lipid bilayer (20 molar equivalence of 9:1 DMPC/POPS) by 1 min (blue) and 20 mins (red).

ANTIBACTERIAL ACTIVITY OF CYCLIC D, L-α-PEPTIDES

In the presence of various microbial, bacterial, and other cellular systems, self-assembling cyclic D, L-α-peptides have shown the ability to disrupt membrane potentials, leading rapidly to cell death.[13] The activity and selectivity of this behavior is strongly dependent upon the amino acid sequence of the cyclic D, L-α-peptide units (Table 1). Minimum inhibitory concentration assays have determined the relative activity of respective derivatives, and fluorescence depolarization studies have indicated that such activity coincides with disruption of the transmembra potential gradient.

TABLE 1. Minimum Inhibitory concentration (µg/mL) vs the indicated bacteria strain.

Peptide*	MRSA	E. coli
c[KQRWLWLW]	6	80
c[KSKWLWLW]	5	40
c[KKKWLWLW]	7	80
c[KXKRWLWR]	25	> 40
c[QXHRWLWK]	12.5	> 40

*Single letter codes for amino acids and shorthand representation of cyclic peptide sequences are used. The brackets indicate cyc structure and underlining represents D amino acid residues; X = pyrene-CH_2CO—Lys.

ACKNOWLEDGMENTS

The authors thank the Skaggs Institute for Chemical Biology and the National Institutes of Health (USA) financial support.

REFERENCES

1. Ghadiri, M. R., Granja, J. R., Milligan, R. A., Mcree, D. E., and Khazanovich, N., *Nature* 366 (1993) 324-327.
2. Ghadiri, M. R.; Granja, J. R., and Buehler, L. K., *Nature* 369 (1994) 301-304.
3. Granja, J. R. and Ghadiri, M. R., *J. Am. Chem. Soc.* 116 (1994) 10785-10786.
4. Ghadiri, M. R.; Kobayashi, K., Granja, J. R., Chadha, R. K., and Mcree, D. E., *Angew. Chem., Int. Ed. Engl* 34 (1995) 95.
5. Kobayashi, K., Granja, J. R., and Ghadiri, M. R., *Angew. Chem., Int. Ed Engl.* 34 (1995) 95-98.
6. Hartgerink, J. D., Granja, J. R., Milligan, R. A., and Ghadiri, M. R., *J. Am Chem. Soc.* 118 (1996) 43-50.
7. Motesharei, K. and Ghadiri, M. R., *J. Am. Chem. Soc.* 119 (1997) 11306-11312.
8. Clark, T. D., Buriak, J. M., Kobayashi, K., Isler, M. P., Mcree, D. E., Ghadiri, and M. R., *J. Am. Chem. Soc.* 102 (199 8949-8962.
9. Kim, H. S., Hartgerink, J. D., andGhadiri, M. R., *J. Am. Chem. Soc.* 120 (1998) 4417-4424.
10. Hartgerink, J. D.; Clark, T. D.; Ghadiri, M. R., *Chem. Eur. J.* 4 (1998) 1367-1372.
11. Rapaport, H., Kim, H. S., Kjaer, K., Howes, P. B., Cohen, S., Als-Nielsen, J., Ghadiri, M. R., Leiserowitz, L., and Laha M., *J. Am. Chem. Soc.* 121 (1999) 1186-1191.
12. Bong, D. T., Clark, T. D., Granja, J. R., and Ghadiri, M. R., *Angew. Chem., Int. Ed. Engl.* 40 (2001) 988-1011.
13. Fernandez-Lopez, S., Kim, H. S., Choi, E. C., Delgado, M., Granja, J. R., Khasanov, A., Kraehenbuehl, K., Long, (Weinberger, D. A., Wilcoxen, K., and Ghadiri, M. R., *Nature* 412 (2001) 452-455.
14. Sanchez-Quesada, J., Kim, H. S., and Ghadiri, M. R., *Angew. Chem., Int. Ed. Engl.* 40 (2001) 2503-2506.
15. Khazanovich, N., Granja, J. R., McRee, D. E., Milligan, R. A., and Ghadiri, M. R., *J. Am. Chem. Soc.* 116 (1994) 6011 - 6012.
16. Forster, T. *Angew. Chem., Int. Ed. Engl.* 8 (1969) 333-343.

ELECTRONIC STRUCTURE

Estimations of Carrier Mobility and Trapped-Carrier Density of States for Microcrystalline Silicon Semiconductors from Analysis of Field Dependent Steady-State Photocarrier Grating Measurements

R. I. Badran

Physics Department, Faculty of Science, King Abdulaziz University, Jeddah, Saudi Arabia
rbadran@kau.edu.sa

Abstract. The drift mobility for electrons and holes, small-signal mobility lifetime product, trapped-carrier density of states in addition to other correlated physical parameters of microcrystalline silicon sample are estimated from the analysis of field-dependent experimental data using steady-state photocarrier grating technique. The filed-dependent experimental data at room temperature of the sample which was prepared by hot-wire chemical vapor deposition (HWCVD) technique, are analyzed using different approaches based on small-signal photocurrent. The exploitation of the electric-field dependence in these approaches is correlating the photoelectronic properties, which are demonstrated by the transport parameters, to the trapped charge density which is usually not easily accessible. This may also justify the enhanced relationship between the minority carrier mobility-lifetime product and trapped charge density and then the sub-gap absorption in the sample under study.

Keywords: (Amorphous semiconductors; glasses), (Electronic transport phenomena in thin films), (Charge carriers: generation, recombination, lifetime, trapping, mean free paths), (Photoconduction and photovoltaic effects).
PACS: 73.61.Jc; 73.50.h; 73.59. Gr; 73.50.Pz

INTRODUCTION

The detection of photocarrier grating amplitude, in a steady-state photocarrier grating technique (SSPG), is a good probe for determining the diffusion length [1-9]. The measurement of small signal conductivity in SSPG depends on the grating period Λ which, in turn, varies as the angle of incidence of two laser beams of different intensities. The photocurrent density measured, in an a-Si:H sample, either when the two laser beams interfere to generate gratings which caused a non-uniform spatial distribution of carrier densities that leads to an ac photocurrent called J_{coh} or when the beams do not interfere where the generated carrier density and then the photocurrent, J_{incoh}, are due to a pure bias illumination. The measured ratio ($\beta = J_{coh}/J_{incoh}$) and its relation to the grating period give the estimated observed value of ambipolar diffusion length L_{amb} [1-3]. Therefore the photoconductivity of a small spatial modulation (or grating) of photocarrier concentration amplitude depends on L_{amb}. The excitation of photocarriers from trap states creates space charge effects in the sample. The generation of the space charges is controlled by both mobile and trapped carriers. The tendency of generating space charges becomes high when the most mobile carriers (electrons) move away faster than the less mobile carriers (holes). The change in the external field from low values to high values results consequently in a change from diffusion-dominated to drift-dominated transport process. Several theoretical studies based on small signal approximations were used to show the effect of varying electric field on β [4-9].

CP929, *Nanotechnology and Its Applications, First Sharjah International Conference*
edited by Y. I. Salamin, N. M. Hamdan, H. Al-Awadhi, N. M. Jisrawi, and N. Tabet
© 2007 American Institute of Physics 978-0-7354-0439-7/07/$23.00

This paper concentrates on the investigation of the field dependence of the coefficient β on the transport parameters by studying the experimental data for microcrystalline silicon thin film sample, using the approaches of Abel et. al. [4], Li [5] and Hattori et. al. [6]. The application of these approaches on such experimental data may enable us to extract important information on the electronic properties (like drift mobility for electrons and holes small signal mobility-lifetime product of charge carriers and recombination lifetime) and to estimate the trapped carrier density. The diffusion length which was determined from fitting the experimental data in the low-field diffusion regime was employed to fit the experimental data for the wide range of electric fields [7]. We found that this assumption can be adopted within the illumination level of the conducted SSPG experiment for the μc-Si:H sample which corresponds to uniform generation rate of 3.8×10^{20} $cm^{-3}s^{-1}$. The sample was prepared by HWCVD technique in ultra-high vacuum system [10].

RESULTS AND ANALYSIS

A FORTRAN program developed for each approach using the necessary adjustable parameters in wide range of electric field values and different grating periods, is used. Each of the used adjustable parameters is changed within a range of reasonable values based on the physical grounds. Each choice of parameters gives us the calculated result of β which is compared to that of corresponding experimental data. However, each program allows us to iterate all possible values of combination of parameters and calculate the χ^2 as an indicator to get the best fit to experiment. The best choice of parameters is taken when the value of χ^2 is minimum. In the three approaches of Abel et al, Li and Hattori et al, the deviation of theoretical results from experimental data is taken within an average of almost 10% of all experimental data at which the value of χ^2 is minimum.

Results Due to Application of Abel *et al.* Approach

In order to fit the field dependence experimental data of β, using Abel et al approach [4], the value of $(\mu\tau)_p$ extracted from the low field fit and $(\mu\tau)_n$ taken from the experimental data are employed as input data, in a FORTRAN computer program [8]. Thus, only one adjustable parameter is used to fit the experimental data, here, namely, $\mu_n \tau_{rel}^{eff}$ and γ is given a value 0.61. Figure 1 shows the variation of the electric field dependence of β for the μc-Si:H sample at 295K.

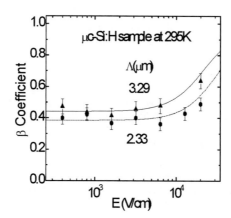

FIGURE 1. The coefficient β versus electric field for the μc-Si:H sample at 295K. Both theoretical results (lines) and experimental data (symbols) are taken at two grating periods. The theoretical results are found applying Abel et. al. approach [4].

Here, the trapped carrier density N_t can be obtained from the relation between the total carrier density N_o [free plus localized] and $\mu_n \tau_{rel}^{eff}$ is approximately equal to $\varepsilon_r \varepsilon_o / e N_o$, by assuming that $N_o \approx N_t$ i.e. the trapped carrier density is much higher than the free carrier density. The list of extracted parameters and the χ^2 values are presented in Table 1.

Results Due to Application of Li Approach

The adjustable parameters a, μ_n, μ_p and g_1/N_o are used in another FORTRAN program [8] to fit the same experimental data. The theoretical results together with the experimental data are shown in Figure 2. The fitting of all experimental data of field-dependent β started with values of μ_n/μ_p (or b) and τ/τ_{diel} (or a) near to or greater than unity in order to avoid breaking the linearization condition for a finite excitation grating [5, 8]. However, the value of L_{amb} extracted from low field fit and inserted as input data in the field dependence fit, is found reasonable at the chosen values of a. The ratio I_2/I_1 is given a value of 0.06 and $\tau_{diel} = \varepsilon_o \varepsilon_r / \sigma_{ph}$ is used. Table 1 lists all adjustable parameters and χ^2 values. Here, the parameters μ_p and μ_p exhibit opposite effects on β at both low and high values of electric field.

Results Due to Application of Hattori *et al.* Approach

Reasonable fits are obtained to experimental data using five adjustable parameters, namely, μ_n, μ_p, μ'_n, μ'_p and τ/τ_{diel} using a third FORTRAN code [9]. The fits to the experimental data of the coefficient β versus applied electric field at room temperature are presented for μc-Si:H sample in figure 3. The values of $\gamma_0 \approx 1$, $\gamma = 0.61$ and $\tau_{diel} = \varepsilon_o \varepsilon_r / \sigma_{ph}$ previously used in Li approach, are also employed here. The value of a is same as that obtained from Li fittings. The fitting parameters μ_n and μ_p are found different from those obtained from Li approach, as listed in Table 1. Again the parameters μ_p and μ_p exhibit opposite effects on β at both low and high values of electric field.

The small-signal electron drift mobility μ'_n is found more sensitive than the hole drift mobility μ'_p in fitting the experimental data. This is reflected in the tolerance in the value of μ'_n of $\pm 0.2 \times 10^{-3}$ cm^{2}V^{-1}s^{-1} while the value of μ'_p is almost invariant.

A reasonable fit could not be reached unless δ is close to unity at the given generation rates of 3.8×10^{20} $cm^{-3}s^{-1}$. This assumption means that a larger value of hole lifetime τ'_p than that of the electron lifetime τ'_n is discernible.

Our estimation to hole mobility values at room temperature disagree by two orders of magnitude with those measured by the transient photocurrent technique [11] and three order of magnitude by those found by the field-effect hole mobility measurement [12], but they agree with those measured by TOF experiment [13]. Moreover, the values of electron mobility differ by almost one order of magnitude to that measured by TOF experiment [13]. Our results also show that $\mu'_n \neq \mu_n$ which is an indication that both types of mobility depend on concentration.

Hattori *et al.* approach gives the least χ^2 values among the three approaches.

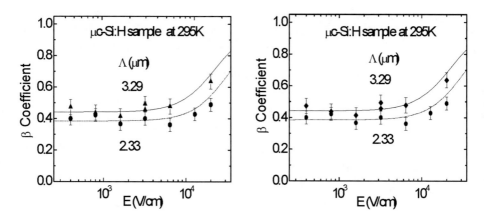

FIGURE 2. The same as in figure 1, but (left panel) from applying Li approach [5] and (right panel) from applying Hattori *et. al.* approach [6].

TABLE 1. List of parameters obtained from the three different approaches as the theoretical results compared to the experimental data of electric field dependence of β for the microcrystalline silicon sample at grating period of 3290 nm and at temperature of 295K. Here δ=1, and η=0 are chosen due to Hattori et. al. [6, 9]. The parameter γ=0.61 for all approaches. The values of χ^2 for all approaches are also listed.

	Li	Hattori	Abel
$a=\tau/\tau_{diel}$	0.51	0.51	-
$\mu_n \times 10^{-2}$ (cm^{2}V^{-1}s^{-1})	0.91	1.26	-
$\mu_p \times 10^{-2}$ (cm^{2}V^{-1}s^{-1})	0.4	0.59	-
$\mu'_n \times 10^{-3}$ (cm^{2}V^{-1}s^{-1})	-	1.88	-
$\mu'_p \times 10^{-3}$ (cm^{2}V^{-1}s^{-1})	-	1.26	-
$g_l/N_o \times 10^5$ (s^{-1})	1.4	-	-
$\mu_n \tau^{eff}_{diel} \times 10^{-8}$ (cm^{2}V^{-1})	-	-	0.24
$(\mu'_n+\mu'_p)\ \tau' \times 10^{-11}$(cm^{2}V^{-1})	4.99	4.95	-
$N_o \times 10^{14}$ (cm^{3}eV^{-1})	3.46	3.35	980.5
χ^2	6.34	0.31	10.5

The formula for the carriers charge density $N_o \approx N_{ph} \approx \tau G_o/\gamma$ due to Li approach[5], is adopted and the value of N_o obtained seems to have same value as that calculated using the same approximate formula but using τ' obtained from fit due to Hattori *et al.* approach [6] instead of τ. Moreover, the calculated value of N_o, due to Abel *et. al.* approach, is in agreement with that obtained by Reynolds *et. al.* [14] Finally, the values obtained, here, for DOS can be ascribed, at most, to the density of deep defects in our sample. These values can also be compared to those of others [13, 15]. The values of N_o in Table 1 together with the value of $(\mu\tau)_p$ may be considered as evidence for pronounced correlation between the trapped charge density and the minority carrier mobility-lifetime product. The increase in sub-gap absorption is consistent with the much larger trapped charge density revealed in the field-dependent SSPG experiments.

CONCLUSION

It is worth noting that the investigation of the hydrogenated microcrystalline semiconductor thin film, grown by hot wire chemical vapor deposition technique using both the measurements and the SSPG theory gives us important information on the photoelectronic properties of this type of samples. The use of SSPG theory comprises three different approaches based on the small-signal photocurrent and involves the electric-field dependence in such technique. The experimental data obtained from SSPG technique that cover a wide range of applied electric field values, probe the whole transition region between the diffusion and drift-dominated transport. The analysis of these experimental data leads to the extraction of the information on carriers transport in addition to the ambipolar diffusion length. The ambipolar diffusion length is determined from fitting the experimental data in the low-field diffusion regime, and then employed in the fit of the experimental data for the whole range of electric fields. The comparison of transport parameters among the three approaches facilities the task of fitting experimental data and open "the door" for a self-contained analysis. This also has the merit of enabling us to estimate several important parameters such as small-signal mobilities of both types of carriers and lifetime. Moreover, the small-signal mobility lifetime product which is an important characteristic parameter in semiconductors, is estimated. The exploitation of the electric-field dependence in the three approaches is correlating the photoelectronic properties, which are demonstrated by the transport parameters, to the trapped charge density which is usually not easily accessible. This may also justify the enhanced relationship between the minority carrier mobility-lifetime product and trapped charge density and then the sub-gap absorption in the sample under study.

Our theoretical investigation demonstrates the weakness in the assumption of local charge neutrality and ambipolarity restriction. It also shows that mobility must be concentration dependent parameter. The extracted transport parameters are found sensitive in the transition region that links the lifetime regime where the ambipolar transport takes place, with the relaxation time where the space-charge effects and bipolar transport become significant. We reckon that the analyses of experimental data of electric field dependence, due to the approaches of Abel *et al.*, and Hattori *et al.*, are more recommended than that of Li approach. Moreover, a well-established formula for the density of states without any prerequisite assumptions from SSPG theory becomes highly necessary for microcrystalline semiconductors. Such formula may allow for more precise determination of DOS from SSPG technique.

Overall, the agreement between the calculations and experimental data is considered reasonable in view of the approximations involved in using the three adopted approaches.

ACKNOWLEDGMENTS

The author would like to thank Dr. Rudi Brueggemann, University of Oldenburg, H. Brummack and K. Bruhne, Institut für Physikalische Elektronik, Universität Stuttgart, for providing the experimental data and the samples.

REFERENCES

1. D. Ritter, K. Weiser and E. Zeldov, J. of App. Phy. **62** (1987) 4563.
2. D. Ritter, E. Zeldov and K. Weiser, Appl. Phys. Lett. **49** (1986) 791.
3. D. Ritter, K. Weiser and E. Zeldov, Phys. Rev. B, **38** (1988) 8296.
4. C. D. Abel, G. H. Bauer and W. H. Bloss, Phil. Mag. *B*, **72** (1995) 551.
5. Y. M. Li, Phys. Rev. B, **42** (1990) 9025.
6. K. Hattori, H. Okamoto and Y. Hamakawa, Phys. Rev. B, **45** (1992) 1126.
7. R. Brueggemann and R. I. Badran,. Mat. Res. Soc. Symp. San Francisco-USA, Vol. **808** (2004) A9.7.1
8. R. I. Badran, J. Mat. Sci.: Materials to electronics, **18** (2007) 405.
9. R. I. Badran and N. Al-Awaad, J. optoelectron. Adv. Mat. **8** (2006) 1466.
10. R. Brueggemann, University of Oldenburg, Germany (Private Communication 2004).
11. T. Dylla, F. Finger and E. A. Schiff, Appl. Phys. Lett. **87** (2005) 0321103
12. L. C. Cheng and S. Wagner, Appl. Phys. Lett. **80** (2002) 440.
13. J. P. Kleider, M. Gauthier, C. Longeaud, D. Roy, O. Saadane and R. Brueggemann,Thin Solid films **403-404** (2002) 188.
14. S. Reynolds, V. Smirnov, F. Finger, C. Main, R. Carius, J. Optoelect. Adv. Mat. **7** (2005) 91.
15. J. P. Kleider, C. Longeaud, M. Gauthier, M. Meaudre, R. Meaudre, R. Butte, S.Vignoli, P. Roca, and I. Cabarrocas, Appl. Phys. Lett. **75** (1999) 3351.

APPLICATIONS

Improving the Erosion Resistance of Electrical Insulating Materials Using Nano Fillers

A. El-Hag[1], S. Ul-Haq[2], S. Jayaram[2] and E. Cherney[2]

[1] *American University of Sharjah, Sharjah, United Arab Emirates*
[2] *University of Waterloo, Waterloo, Canada*
aelhag@aus.edu

Abstract. The paper presents the experimental results obtained to test the effect of nano-fillers on the aging performance of silicone rubber for outdoor applications and enamelled wire for motor insulation. The erosion resistance of silicone rubber (SIR) filled with 12 nm size fumed silica is compared to those filled with 5 μm size silica filler using the ASTM 2303 Inclined Plane Tracking and Erosion Test. The erosion resistance of the SIR materials increased with increasing percentage of the fillers, and it was observed that 10% by weight of nano-filled SIR gives a performance that is similar to that obtained with 50% by weight of micro-filled SIR. The paper discusses the possible reasons for the improvement in the erosion resistance of nano-filled silicone composites using different material analysis techniques like Thermo Gravimetric Analysis (TGA), and Scanning Electron Microscopy (SEM). Also, the effect of using different nano fillers like alumina, fumed silica and titanium oxide on the erosion resistance of enamel wire insulating material subjected to different electrical stresses will be addresses. Surface roughness is used to evaluate the effect of different nano-fillers on the erosion resistance of enamel wire insulation.

Keywords: Aging, electrical insulation, nano-dielectrics, nano fillers.
PACS: 81.07.Wx

INTRODUCTION

In the past few years, a great deal of attention has been given to the application of nano-dielectrics in the field of electrical insulating materials. Nano-dielectrics are a class of materials containing at least one phase at the nanometer scale [1]. It has been reported that the use of nano-particles in the matrix of polymeric materials can greatly improve the thermal, mechanical and electrical properties of polymeric composites [2-4]. Researchers, scientists, and practitioners across almost the entire spectrum of disciplines, are doing efforts by exploring and developing science and engineering at the nano-level. Among these disciplines, the progress in the field of polymer sciences has demonstrated that the addition of nanofillers represents a very attractive route to upgrade and diversify material properties. There is a broad range of nanofiller types, which has led to nanopowders with different morphologies (size and shape, cluster composition, and dispersion) that are currently being used in different electrical insulating materials.

Various studies have been made comparing the performance of nano- to micro-particle filled composites. Compared to micro-fillers, nano-fillers have been reported to mitigate more efficiently the space charge formation in polymeric materials [4-6]. It has also been reported that the use of nano-fillers improves the corona resistance of polyimide films [7]. In addition, the use of nano-particles of zinc oxide in low density polyethylene has shown a smaller decrease in the resistivity of the polymeric matrix compared to micro-fillers [8]. The time-to-failure of cable material due to electrical treeing in epoxy resin has been increased using nano-particle fillers compared to micro-particle fillers [10]. Hudon et al [11] have reported that the nanofillers in magnet wire enamel are essential for a variety of reasons, including discharge

CP929, *Nanotechnology and Its Applications, First Sharjah International Conference*
edited by Y. I. Salamin, N. M. Hamdan, H. Al-Awadhi, N. M. Jisrawi, and N. Tabet
© 2007 American Institute of Physics 978-0-7354-0439-7/07/$23.00

resistance, matching the coefficient of the thermal expansion, thermal conductivity enhancement, mechanical reinforcement, and abrasion resistance.

In this study the authors will investigate the influence of nanofillers in the aging performance of two different electrical insulating materials, i.e. silicone rubber for outdoor applications and polyimide material with and without nanofillers for magnet wire applications.

MATERIALS AND METHODS

Aging of Silicone Rubber Material

The test procedure followed the ASTM 2303 standard [12], with an initial voltage of 2.0 kV and a constant contaminant flow rate of 0.15 ml/min for four hours. The voltage was increased at a rate of 250V/hour. At the end of the test, the samples were taken from the test bay and the eroded volume was estimated by filling the eroded volume with a soft putty of known density and then the eroded volume was calculated. The base rubber (elastomer) used was a two-component room temperature vulcanized (RTV) SIR material which contained no inorganic filler of any type. The filler type/concentration used is presented in Table 1. Figure 1 shows the schematic of inclined plane experimental setup.

TABLE 1. Chemical composition of the tested samples in IPT test.

Filler Type and median size	Concentration of filler by weight (%)
12 nm fumed silica	5
12 nm fumed silica	10
5 μm ground silica	10
5 μm ground silica	30
5 μm ground silica	50

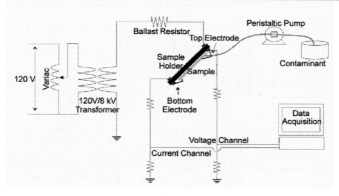

FIGURE 1. IPT Experimental setup.

Aging of Polyimide Material

The aging performance of six different samples (S_1 to S_6) of magnet wire was investigated. While samples S_1-S_3 are commercial samples, samples S_4-S_6 are built following NEMA MW-1000 specifications [13]. The details of the six samples are shown in Table 2.

For the insulation aging tests to be representative of the voltages that result from medium voltage (2.3-6.6 kV) PWM drives, a lab-built solid-state based PWM-VSC is utilized. For the circuit to safely and reliably operate at higher voltages it utilizes a chain of insulated gate bipolar transistor (IGBT) switches connected in series. The control circuit consists of the PIC microcontroller with its own custom program used to generate the appropriate gating signals. This generator can reproduce the phase-to-ground voltage of a two-level PWM drive. Figures 2 and 3 show a single line diagram of the test setup along with a PWM waveform, with a fundamental frequency of 60 Hz and switching frequency of 1.25 kHz, respectively 14].

TABLE 2. Chemical composition of the six tested samples.

Sample no.	Nano Filler Type
S_1	N/A
S_2	Alumina
S_3	Alumina
S_4	Fumed silica
S_5	Titanium oxide
S_6	Alumina

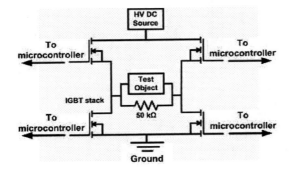

FIGURE 2. Single line diagram of test setup, used for pulse aging [14].

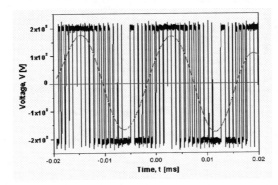

FIGURE 3. PWM waveform from a PWM generator used for testing of magnet wire specimens (S_1 to S_6).

211

RESULTS AND DISCUSSIONS

Silicone Rubber Material

The eroded volumes for all the tested samples were measured and are depicted in Fig. 4. It is evident from Fig. 4 that as the filler concentration increased, the eroded volume decreased. Also, the weight loss for a 10% nano-size silica filled SIR is comparable to a 50% micro-size silica filled SIR.

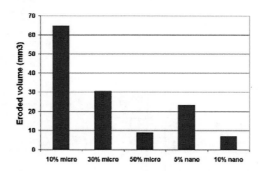

FIGURE 4. Comparison between micro- nano-size silica filled composites in terms of eroded volume.

The TGA analysis was conducted on both 10% nano- and micro- size silica filled SIR, Figs. 5. Comparing the TGA results of nano- and micro- size silica filled SIR do not agree with the IPT results. The TGA analysis show that both the 10% nano- and micro- size silica filled SIR materials show a reduction in weight of about 30%. On the other hand, the eroded volume for 10% nano-size silica filled SIR is about one sixth of the eroded volume of the micro-size silica filled SIR.

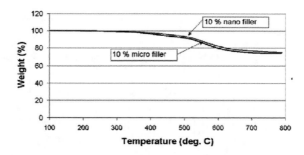

FIGURE 5. TGA Analysis for 10% nano and micro filled silicone rubber samples.

To further understand the difference between nano- and micro-size silica filled SIR, a detailed scanning electron microscopy (SEM) investigation was conducted on both the damaged 10% nano- and micro-size silica filled SIR samples as shown in Fig. 6. There is a significant difference between the surfaces of the damaged nano- and micro-size silica filled SIR. The fumed silica tended to agglomerate during the dry band arcing, forming a silica-like layer, Figure 6-b. The formation of this type of a layer was not apparent in the micro-size silica filled samples, as shown in Figure 6-

. It can therefore be stated that the formation of silica like layer with nano-size silica filled composites aids to resist further degradation of SIR during dry band arcing.

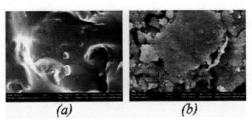

(a)   *(b)*

FIGURE 6. SEM Images for different SIR filled materials; (a) Damaged 10% micro-size silica filled, (b) Damaged 10% nano-size silica filled.

Polyimide Material

The magnet specimens (S_1 to S_6), examined for the PD resistance, are exposed to a stress of 100 kV_p/mm for 1 h under PWM waveforms. Table 3 lists the average, mean, and standard deviation of the results. It is evident from Table 3 that the higher average surface roughness of approximately 1195 nm with a standard deviation of 134.5 nm is observed for enameled wire S_1 without nanofillers. Similarly, for wires S_2 and S_3, which are commercially available PD resistant wires, reveal similar surface erosion, after comparing both un-aged and aged samples. The average surface roughness is around 327 nm for S_2 and 313 nm for S_3, along with the piling up of nanoparticles, to reduce the PD attack. Also, the standard deviation values are lower, compared with wire S_1, which are 63 and 45.5 nm, respectively.

In magnet wires S_4, S_5, and S_6, filled with fumed silica (SiO_2), TiO_2, and Al_2O_3, respectively, the surface erosion, due to the PD, is comparatively lower. In case magnet wires S_4 and S_5, the PD erosion seems to be around 240 nm and 280 nm, respectively. In addition, in both cases, the scatter in the data is smaller; whereas, in comparison, S_6 shows surface erosion relatively higher in the range of 322 nm with a standard deviation of 52.5 nm.

Similarly to what has been noticed for SIR, nanofillers prevent further degradation for the magnet wire samples by the formation of a shield like layer on its surface. This can be explained schematically using Fig. 7-a and SEM images for S_5 is shown to support this mechanism.

For both SIR and polyimide materials it has been found that it is extremely difficult to increase the nanofiller concentration. In case of magnet wire coatings, when the concentration of filler exceeds 1% as the enamel coating becomes rough and brittle and unsuitable for magnet wire as depicted in Fig. 8. Also, it was extremely difficult to mix more than 10% of fumed silica with SIR material due to the increase of the polymer viscosity. This may be due to agglomeration of particles, which needs further investigation.

TABLE 3. Average, mean, and standard deviation of the surface roughness in magnet wire coatings subjected to PWM waveforms at 100 kV_p/mm.

Sample no.	Surface Roughness (nm)		
	Average	Mean	SD
S1	1195	1166	134.5
S2	327	322	63.0
S3	313	320	45.5
S4	240	230	35.5
S5	280	270	38.8
S6	322	313	52.5

a) Schematic Illustration.

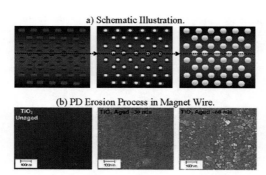

(b) PD Erosion Process in Magnet Wire.

FIGURE 7. Comparison of schematic illustration of the erosion process on nanoparticles surface in magnet wire with a constant stress of 70 kV$_p$/mm.

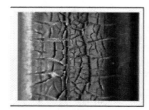

FIGURE 8. Magnet wire with more than 1% fumed silica.

CONCLUSIONS

This research has shown that the erosion resistance of both SIR and enameled wire has improved due to the application of nanofillers. It has been found that the formation of a shield layer on the material surface prevent any further degradation.

REFERENCES

1. M.F. Frechette, M. Trudea, H.D. Alamdari and S. Boily, "Introductory remarks on nanodielectrics", IEEE Transaction on Dielectrics and Electrical Insulation, Vol. 11, No. 5, October 2004, pp. 808-818.
2. P.C. Irwin, Y. Cao, A. Bansal and L.S. Schadler, "Thermal and mechanical properties of polyimide nanocomposites", IEEE CEIDP-2003, pp. 120-123, Albuquerque, 2003.
3. Y. Cao, and P.C. Irwin, "The electrical conduction in polyimide nanocomposites", IEEE CEIDP-2003, pp. 116-119 Albuquerque, 2003.
4. T. Imai, Y. Hirano, H. Hirai, S. Kojima and T. Shimizu, "Preparation and properties of epoxy-organically modified layered silicate nanocomposites", IEEE International Symposium on Electrical Insulation, pp. 379 – 383, Boston, 2002.
5. J. K. Nelson, Y. Hu and J. Thiticharoenpong, "Electrical properties of TiO2 Nanocomposites" IEEE CEIDP-2003, pp. 719-722 Albuquerque, 2003.
6. Yi Yin, Jiong Chen, Jingning Yang, Dengming Xiao, Demin Tu, Rui Yin and Hongjin Qian, "Effect of space charge in nanocomposites of LDPE/TiO2", Proceedings of the 7th International Conference on Properties and Application of Dielectric Materials, pp. 913-916, Nagoya, 2003.

. Z. Peihong, Z. Weiguo, L Yan, F. Yong and L. Qingquan, "Study on corona resistance of polyimide nano inorganic composites", Proceedings of the 7th International Conference on Properties and Application of Dielectric Materials, pp. 1138-1141, Nagoya, 2003.

. Y. Cao, P. Irwin and K. Younsi, "The future of Nanodielectrics in the electrical power industry", IEEE Transaction on Dielectrics and Electrical Insulation, Vol. 11, No. 5, October 2004, pp. 797-807.

. Kozako, M., Fuse, N., Ohki, Y., Okamoto, T. & Tanaka, T., "Surface degradation of polyamide nanocomposites caused by partial discharges using IEC (b) electrodes" IEEE Transaction on Dielectrics and Electrical Insulation, Vol. 11, No. 5, October 2004, pp. 833- 839.

0. H. Z. Ding and R. Varlow, "Effect of nano-fillers on electrical treeing in epoxy resin subjected to AC voltage", IEEE CEIDP-2004, pp. 332-335, Boulder, 2004.

1. C. Hudon, N. Amyot and D. Jean, "Long term behavior of corona resistant insulation compared to standard insulation of magnet wire", IEEE-ISEI, pp. 13-16, 2000.

2. ASTM-D2303 "Standard Test Method for Liquid-Contaminant, Inclined-Plane Tracking and Erosion of Insulating Materials".

3. NEMA Standard Publication, MW 1000-2003, Magnet Wire, 2003.

4. F. P. Espino-Cortes, Y. Montasser, S. H. Jayaram, E. A. Cherney and L. C. Simon, "Study of stress grading systems working under fast rise time pulses", IEEE International Symposium on Electrical Insulation (ISEI), pp. 380-383, 2006.

215

Physical Properties of Polyester Fabrics Treated with Nano, Micro and Macro Emulsion Silicones

M. Parvinzadeh[a] and R. Hajiraissi[b]

[a] Color Control and Reproduction Department, Institute for Colorants, Paint and Coating, Tehran, Iran
[b] Islamic Azad University, Tehran South Branch, Tehran, Iran
hajiraissi@yahoo.com

Abstract. The processing of textile to achieve a particular handle is one of the most important aspects of finishing technology. Fabrics softeners are liquid composition added to washing machines during the rinse cycle to make clothes feel better to the touch. The first fabric softeners were developed by the textile industry during the early twentieth century. In this research polyester fabrics were treated with nano, micro and macro emulsion silicone softeners. Some of the physical properties of the treated fabric samples are discussed. The drapeability of treated samples was improved after treatment with nano silicone softeners. The colorimetric measurement of softener-treated fabrics is evaluated with a reflectance spectrophotometer. Moisture regain of treated samples is increased due to coating of silicone softeners. There is some increase in the weight of softener-treated samples. Samples treated with nano emulsion silicones gave better results compared to micro- and macro-emulsion treated ones.

Keywords: Emulsion, Drape length, Colorimetric properties, polyester.
PACS: 68.37Hk; 81.07.-b

INTRODUCTION

Polyester is currently defined as long chain polymers chemically composed of at least 85 percent by weight of an ester and a di-hydric alcohol and a terephthalic acid. The name "polyester" refers to the linkage of several monomer (esters) within the fiber. Esters are formed when alcohol reacts with a carboxylic acid. Polyester is often used in outwear because of its high tenacity and durability. It is a strong fiber and consequently can withstand strong and repetitive movements. Polyesters is used as in pants, shirts, suits and bed sheets either by itself or as blend because of its wrinkle-resistant property and its ability to retain its shape. Polyester has industrial uses as well, such as carpets, filters, synthetic artery replacements, ropes and films.

Finishing is the final series of operations that produce finished textile fabric from greige goods. Finishing operations are predominantly wet operations requiring large amounts of thermal energy for water heating and drying. Woven greige goods require some additional steps prior to dyeing, as compared to knit goods. As the first step in finishing woven goods, singeing burns protruding fibers by passing the fabrics over an open flame or heated plates to produce a cleaner fabric and enhance future operations. Hand or handle are usually used to describe fabric drapes around an object or feels to the touch. Softeners designed to impart a soft mellowness to the fabric. They also change some of textile properties and they often have a multifunctional nature. They improve abrasion and soiling resistance, crease recovery, static protection, increase tearing strength and stretch, reduce pilling, color fastness, moisture absorbency, flammability, sewing thread breakage and needle cutting when the garment is sewn. They are classified according to their ionic character and the main classes are: anionic, cationic, nonionic, amphoteric, reactive, and silicone. Silicones were classified as a separate class of man-made polymer more than 50 years ago. Two commercial silicone softeners are

CP929, *Nanotechnology and Its Applications, First Sharjah International Conference*
edited by Y. I. Salamin, N. M. Hamdan, H. Al-Awadhi, N. M. Jisrawi, and N. Tabet
© 2007 American Institute of Physics 978-0-7354-0439-7/07/$23.00

methylsilicones and amino silicones. Silicones have different effects on fibrous substrates. The most typical properties imparted to fibers, due to the above mentioned extreme flexibility of their backbone and to their low surface energy, are:

Highly reduced fiber friction results in soft yarn and fabric handle as well as in improved human hair combability (the ease of hair combing). Hydrophobic fiber surface leads to water repellency up to very high levels.

The performance of softeners is judged by: handle, whiteness/tendency to yellowing, hydrophilic/hydrophobic character i.e. Fabric wettability, shear stability under application conditions and exhaustion from long liquor (if required). Silicones can anticipate a bright future. They are one of the most innovative classes of textile finishing auxiliaries.

In this research, effect of nano, micro and macro emulsion silicones softeners on physical properties of polyester fabrics was investigated.

EXPERIMENTAL

Polyester Fabric was used and properties are shown in Table 1. The weight of samples before and after softener treatment was measured using a digital balance. Moisture regain was assessed according to ASTM method 2654 – 76, and calculated using

$$\text{Moisture Regain} = \frac{w_1 - w}{w_2} \times 100. \tag{1}$$

TABLE 1. Properties of polyester fabric used.

Fabric	Area density (gr/m²)	Number of wraps (1 cm)	Number of wefts (1cm)	Wrap count (Ne_c)	Weft count (Ne_c)	Structure
polyester	156	25	20	40	40	plain

Nonionic detergent from SDL (Shirley Development Limited) was used for scouring. Nano, micro and macro silicone softeners were applied for coating the polyester fabric. Materials were supplied by Eksoy, Turkey. Polyester fabrics were first scoured with 1% nonionic detergent for 30 min at 50°c with the liquor to good ratio of 40:1. Fabrics were then treated with three concentrations of nano, micro and macro emulsion silicone softeners in water including 10, 20 and 30 (gr/lit) at 30°c for 60 min. The PH was maintained at 5 using Acetic acid. The treated fabrics were then dried at 140°c for 3 min. Drape of fabrics at length and width was determined using a Drapometer M003B (SDL technologies) according to ASTM D1388-96. CIELAB color coordinates (L *, a*, b*, C*, h) of the treated samples were evaluated using a Gretagmacbeth Spectrophotometer Color-eye 7000A integrated with an IBM-PC for 10° observer and D65 illuminant.

RESULT AND DISCUSION

Table 2 shows the weight of nano, micro and macro emulsion softeners treated fabrics compared to the untreated one. As can be seen in Table 2, the weight of a sample treated in 10 gr/lit softener solution increased and an increased more in weight was observed with increase in softener's concentration in solution. All the softeners increased the weight of fabric and that doesn't depend on the silicone emulsion size, but less increase was observed for nano emulsion silicone softeners.

Drape length of nano, micro and macro emulsion silicone softeners treated samples in wrap direction is shown in Table 3. As can be seen, 10 gr/lit of softener solution improved drapeability of fabric and decreased the drape length. Increase in softener concentration in solution caused more decrease in drape length. The drape length of nano emulsion silicone treated fabric was less than those treated with micro and macro emulsion silicones.

Table 4 shows moisture regain of softener treated samples. According to the results, all the softeners increased the moisture regain percentage of fabrics but less change was observed for nano emulsion silicone treated samples.

TABLE 2. Weight of fabrics (gr) before and after treatment with nano micro and macro emulsion silicone softeners.

Sample	Quantity of softener in solution(gr/lit)	Fabric weight(gr)
Untreated	--	5
Macro emulsion silicone	10	5.21
	20	5.22
	30	5.26
Micro emulsion silicone	10	5.19
	20	5.23
	30	5.26
Nano emulsion silicone	10	5.19
	20	5.2
	30	5.21

TABLE 3. Drape length (cm) of fabrics before and after treatment with nano, micro and macro emulsion silicone softeners in war direction.

Sample	Quantity of softener in solution(gr/lit)	Drape length(cm)
Untreated	--	6.3
Macro emulsion silicone	10	5.8
	20	5.7
	30	5.5
Micro emulsion silicone	10	5.58
	20	5.5
	30	5.38
Nano emulsion silicone	10	5.56
	20	5.35
	30	5.2

TABLE 4. Moisture regain of fabrics before and after treatment with nano, micro and macro emulsion softeners.

Sample	Quantity of softener in solution(gr/lit)	Moisture regain (%)
Untreated	--	0.5
Macro emulsion silicone	10	3.03
	20	4.04
	30	8.65
Micro emulsion silicon	10	1.98
	20	3.03
	30	4
Nano emulsion silicone	10	1.063
	20	2.083
	30	2.127

Colorimetric properties of silicone-treated fabrics are shown in Table 5. There was no change on lightness (L*) o macro emulsion silicone treated samples. The lightness of nano emulsion silicone softeners increased a little. Accordin to color coordinates, all the softeners caused a little change on color of the samples.

CONCLUSION

PET fabrics were treated with nano, micro and macro emulsion silicone softeners. All the softeners increased the weight of fabrics. Drapeability of nano silicone softeners was more than micro and macro emulsions. More penetration of nano emulsion silicones into the fiber interior worked to improve the drapability of the samples compared to micro and macro emulsion silicones. There was no considerable change on the colorimetric properties of fabrics after treatment with softeners.

TABLE 5. Colorimetric evaluation of fabrics before and after treatment with nano , micro and macro emulsion softeners.

Sample	Quantity of softener in solution(gr/lit)	L*	a*	b*	C*	h
Untreated	--	88.595	3.068	-11.795	12.187	284.579
Macro emulsion silicone	10	88.433	2.949	-11.925	12.284	283.892
	20	88.448	3.132	-12.525	12.911	284.038
	30	88.366	3.134	-12.334	12.726	284.255
Micro emulsion silicon	10	88.468	2.986	-11.999	12.365	283.976
	20	88.661	2.959	-11.682	12.051	284.215
	30	87.694	2.686	-11.569	11.919	283.924
Nano emulsion silicone	10	88.948	3.091	-12.219	12.604	284.194
	20	89.183	3.046	-12071	12.450	284.163
	30	89.113	3.142	-12.222	12.619	284.416

REFERENCES

1. A. Nakamura, *Fiber Science and Technology* (*Science Publishers*, Inc, USA, 17-21, 2000) pp. 17 – 21.
2. R. W. Moncrieff, *Man-Made Fibers* (Halstead Press, New York, New York, 1999).
3. W. S. Perkins, *Textile coloration and finishing* (*Coroliha Academic Press*, North Carolina, 1996).
4. B.Wahle and J. Falkowski, "Softeners in textile processing". Part 1: *An Overview*, Rev. Prog. Color **32**, 118 (2002).
5. P. Habereder and A. Bereck , "Softeners in textile processing". Part 2: *Silicone softeners*, Rev. Prog. Color **32**, 125 (2002).
6. R. Zyschka, "Textile softeners and their tricky application", *Melliand International*, 249-251, 2001.
7. R. Nahta, "Textile softeners", *American Dyestuff Reporter*, 22-26 (August, 1981).
8. P. Habereder and A. Bereck, "Silicone Softeners", Rev. Prog. Color **32**, (2002).
9. M. Parvinzadeh and A. Kiumarsi, "Effects of Softeners on Colorimetric and Fastness Properties of Sulphur Dyed Cotton Fabrics", International Textile Conference, 19-21 (Istanbul, May 2005).

Microscopic Characterization of Silk Fibers Coated with ZnO Nanoparticles

M. H. Ashrafi, [a] A. Kiumarsi, [b] R. Khajavi [a] and M. Parvinzadeh[b]

[a] *Postgraduate Faculty of Islamic Azad University South Tehran Branch, Tehran, Iran*
[b] *Color Control and Color Reproduction Department, Institute for Colorants, Paint and Coating, Tehran, Iran*
meashrafi@yahoo.com

Abstract. For centuries silk had a reputation as a luxurious and sensuous fabric, and been associated with wealth and success. It is one of the oldest textile fibers known to man. It has been used by the Chinese since the 27th century BC. This fiber is sensitive to UV radiation and sun light can damage it. So, designing and modifying silk fibers in such a way that they offer high protection against UV radiation provides some relatively new applications. In this work, ZnO nanoparticles are applied with dip-pad-cure method on silk fiber surface. Scanning Electron Microscope (SEM) was used to indicate the ZnO nanoparticles on silk fiber surface. ZnO nanoparticles are susceptible materials for UV protection textiles.

Keywords: Silk fibers, ZnO, SEM, nanoparticles.
PACS: 81.07.–b

INTRODUCTION

Nanomaterials in the form of dry powder or liquid dispersions are combined with other materials to improve functionality such as: sunscreens and cosmetics, longer lasting tennis balls and lighter, stronger tennis racquets, stain free clothing and mattresses, UV protection clothes, dental bonding agents, burn and wound dressings.

In order to further develop textile application, an increasing demand has arisen for special functionalities of the material. Such functionalities may be achieved by either physical or chemical methods to modify the surface of existing fibrous material. Since the worldwide trend of fiber manufacture is to concentrate on only few standard types, the field of surface modification to impart functions to the textile material has increased. Tecnotessile Co. has developed some tools such as nanotechnology and nano-particles to setup new finishing processes to realize functional textiles [1].

Among numerous traditional and new technologies of surface treatments of textiles and films, the nano-finishing treatment is a powerful tool to fulfill environmental requirements and very specific functions.

Inorganic compounds represent an unexplored field in the textiles industry. Many functionalizations can be achieved using the properties of these materials.

Nanoparticles used in textiles are obtained with sol-gel synthesis; the process is realized in water or organic solvents. The nano-particles possess more surface than micrometer particles and consequently some properties change. A small quantity can obtain great effects. The reduced dimensions enhance the anchoring on the textiles surfaces and it simplifies the formulation of new products. The 100nm particles seem to give the best performances.

The application of nanotechnology to textiles and apparel industry has started only several years ago. The application of nanoparticles compounds with new methodology brings to textiles with enhanced functionalization. Some common nanofinishing of textiles are hydrophilicity, water and oil repellence, anti-bacteria, anti-static, pilling resistance, wrinkle resistance, mechanical resistance and UV-protection.

CP929, *Nanotechnology and Its Applications, First Sharjah International Conference*
edited by Y. I. Salamin, N. M. Hamdan, H. Al-Awadhi, N. M. Jisrawi, and N. Tabet
© 2007 American Institute of Physics 978-0-7354-0439-7/07/$23.00

UV protection is one of the most important functionalization finishing on textiles. Inorganic UV blockers are more preferable to organic ones as they are non-toxic and chemically stable under exposure to both high temperatures and UV. Inorganic UV blockers are usually certain semiconductor oxides such as TiO_2, ZnO, SiO_2 and Al_2O_3. Among these oxides, zinc oxide (ZnO) [2,3] and titanium dioxide (TiO_2) [4-6] are commonly used. It was determined that nano-sized zinc oxide and titanium dioxide were more efficient at absorbing and scattering UV radiation than the conventional size, and therefore block UV radiation more efficiently [2-4]. Rayleigh's scattering theory stated that scattering was strongly depended upon the wavelength and it inversely was proportional to the fourth power. This theory predicts that in order to scatter UV radiation between 200 to 400 nm, the optimum particle size will be between 20 - 40 nm [5]. A thin layer of zinc oxide is formed on the surface of the treated cotton fiber which provides excellent UV-protection [4-7].

This paper gives an overview of surface modification of silk fiber with Dip-pad-cure method only feasible so far on a laboratory scale but which may find broad use if properly developed.

EXPERIMENTAL

The silk was Iranian yarns of 50/2 den, the nonionic detergent used for the scouring of silk yarns, nano ZnO (40% in water) were 60 nm from BYK Co.

The yarns were scoured in 5% nonionic detergent for 15 min at 50°C. The L:G (liquor to good ratio) of the scouring bath was kept at 50:1. A scoured yarn was retained untreated for reference.

The zinc oxide nanoparticles might be applied onto the substrate by dip-pad-cure process to a thick layer. The substrates were dipped in nano ZnO suspension for 5 minute and then padded at nip pressure. The padded silk yarns were air-dried and finally cured at 120 C for 30 min in a pre-heated curing oven, to ensure nanoparticles adhesion to the substrate (silk surface). Then, treated silk yarns were soaked into the ZnO suspension for 10 hours to form zinc oxide nano-crystals on the silk surface. Finally, silk yarns were scoured with nonionic detergent in the same way stated earlier.

The surface of treated silk yarns was investigated using Scanning Electron Microscope (SEM XL30, Philips). The surface of samples was first coated with a thin layer of gold (~10 nm) by Physical Vapor Deposition method (PVD) using a sputter coater (SCDOOS, BAL-TEC). The presence of Zinc as ZnO on silk yarns was also proved with both Energy Dispersive X-ray Microanalysis (EDX) and ash content method.

To measure the ash content samples were dried, weighed and heated at 600°C for 2 hours. The ash residue of each sample was weighed to constant weight.

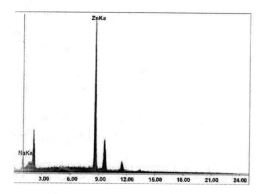

FIGURE 1. EDX ZAF quantification of silk fiber treated with ZnO.

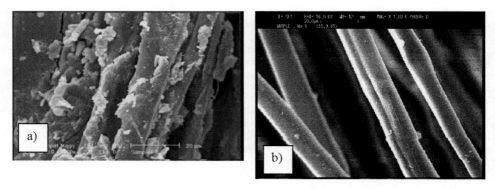

FIGURE 2. A low-magnification SEM image of a) ZnO particles on silk fibers, and b) untreated silk fibers.

FIGURE 3. High-magnification SEM images of ZnO crystals on dip-pad-cure processed silk fibers a) at 5000X, b) 30000X, and c) 40000X magnitude.

RESULTS AND DISCUSSION

It was reported that inorganic oxides such as TiO_2, ZnO, SiO_2 and Al_2O_3 block UV radiation. Among these oxides, zinc oxide (ZnO) and titanium dioxide (TiO_2) are commonly used. It is also believed that nano-sized zinc oxide and titanium dioxide would be more efficient at absorbing and scattering UV radiation than the conventional size, and herefore block UV radiation more efficiently The results of measuring ash contents of non-treated and treated samples, 7.41% and 32.61% respectively, clearly state the formation of nanoparticles of zinc oxide on treated silk fiber surface. EDX analysis also provides clear evidence of zinc as the main inorganic component of samples (Figure 1).

The SEM images were used to investigate the formation of ZnO nanoparticles on silk fiber surface (Figures 2-4). Figure (2-a) shows zinc oxide particles on the surface of treated silk fibers at 1000 magnitude which can be compared to he smoothy surface of untreated silk fibers (Figure 2-b).

The high-magnification SEM images of dip-pad-cure processed silk fibers show ZnO crystals with different diameters ranging from 60 to 100 nanometer formed on the surface of fiber (Fig. 3).

The SEM images of the silk fibers underwent an additional formation process for 10 hours show aggregated ZnO nanoparticles on the surface of the fiber different from previous samples (Figure 4). The size of crystals was also bigger (100 – 140 nm).

 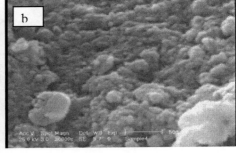

FIGURE 4. SEM images of ZnO crystals on formation processed silk fibers a) at 2000X and b) 30000X magnitude.

The results of this study show that nano ZnO can be coated on silk fiber using conventional dip-pad-cure and formation at ambient temperature and pressure although the method must be progressed.

CONCLUSION

The aim of this study has been to obtain UV protective silk fibers which can be used for people with sensitive skin against UV rays. Inorganic UV blockers such as ZnO are usually certain semiconductor oxides which are commonly used. It can be determined that nano-sized zinc oxide are more efficient at absorbing and scattering UV radiation than the conventional size, and therefore block UV radiation more efficiently. The UV Protective Factor (UPF) of the treated silk fibers will be measured in our future studies.

REFERENCES

1. L. Corsi, S. Nesti, and E. Venturini, "New Finishing Processes to Impart Functional Properties on Textiles Surfaces".
2. M. Saito, Journal of Coated Fabrics **23**, 150 (1993).
3. M. N. Xiong, G. X. Gu, B. You, and L. M. Wu, Journal of Applied Polymer Science **90**, 1923 (2003).
4. J. H. Xin, W. A. Daoud, and Y. Y. Kong, Textile Research Journal **74**, 97 (2004).
5. N. Burniston, C. Bygott, and J. Stratton, Surface Coatings International Part A, 179 (2004).
6. H. Y. Yang, S. K. Zhu, and N. Pan, Journal of Applied Polymer Science **92**, 3201 (2003).
7. W. A. Daoud, and J. H. Xin, Journal of Sol-Gel Science and Technology **29**, 25 (2004).

Preparation of Fragrant Microencapsules and Coating on Textiles

M. H. Shah Jafari[1], M. Parvinzadeh[2] and F. Najafi[2]

[1] *Islamic Azad University - Shahr-e-rey Branch, Iran*
[2] *Institute for Colorants, Paint & Coating, Tehran, Iran*

Abstract. A microcapsule is a small sphere with a uniform wall around it. Microcapsules range in diameter from 1 to 1000 μm. The move by the more developed countries into textiles with new properties and added value, into medical and technical textiles, has encouraged the industry to use microencapsulation process as a means of imparting finishes and properties on textiles which were not possible or cost–effective using other technology. Numerous attempts have been made at adding fragrances directly to fiber and fabrics but all fail to survive after one or two wash cycle. Only through microencapsulation, fragrances are able to remain on a garment during a significant part of its lifetime. This research has tried to prepare microcapsules with poly methyl methacrylate (PMMA) as wall and Rose fragrance as core.

Keywords: Microencapsulation, Microcapsule, Comb Acrylic Copolymer, Fragrance textile.
PACS: PTu.120

INTRODUCTION

Microencapsulation is a process in which tiny particles or droplets are surrounded by a coating to give small capsules. The material inside the microcapsule is referred to as the core, internal phase, or fill, whereas the wall is called a shell, coating, or membrane.

A wide range of materials has been encapsuled as core materials. This includes: solvents, plasticizers, catalysts, colorants, fragrances, detergents, bleaches and so on. Wall Materials may be natural, semisynthetic, or synthetic Microencapsulation methods can be classified into the following groups:

1) Phase separation.
2) Interfacial and in situ polymerization.
3) Spray drying, spray congealing.
4) Solvent evaporation.
5) Coating.

The textile industry has been slow to react to the possibilities of microencapsulation, although by the beginning of the 1990s a few commercial applications were appearing with many more at the research and development stage. As the industry moves into the 21st century the number of commercial applications of microencapsulation in the textile industry continues to grow, particularly in Western Europe, Japan and North America. The move by the more developed countries into textiles with new properties and added value, into medical textiles and technical textiles for example, has encouraged the industry to use microencapsulation processes as a means of imparting finishes and properties on textiles that were not possible or cost-effective using other technology.

In textiles, the major interest in microencapsulation is currently in the application of durable fragrances and skin softeners, insect repellants, dyes, vitamins, antimicrobial agents, phase-change materials, medical applications, polychromic and thermochromic microcapsules, fire retardants, and so on.

CP929, *Nanotechnology and Its Applications, First Sharjah International Conference*
edited by Y. I. Salamin, N. M. Hamdan, H. Al-Awadhi, N. M. Jisrawi, and N. Tabet
© 2007 American Institute of Physics 978-0-7354-0439-7/07/$23.00

The addition of fragrances to textiles has been carried out for many years in the form of fabric conditioners in the wash and during tumble-drying; all are designed to impart a fresh aroma to the textile. However, no matter the quality of the technology used to impart the fragrance, the effect is relatively short-lived. Numerous attempts have been made at adding fragrances directly to fiber and fabrics but all fail to survive one or two wash cycles. Only through microencapsulation, fragrances are able to remain on a garment during a significant part of its lifetime.

MATERIALS AND METHODS

The following materials were used in the study: Rose essence (Gulf Flavours Fragrances, UAE) was selected as the core; and Comb Acrylic Copolymer with Carboxylic groups (DAPCO, Iran) as the shell. Acetone (%98 w/w, Merk, Germany) was used as the solvent; and soft water as an anti-solvent. OAF emulsifier (DAPCO, Iran) was used for preventing the microcapsules to aggregate. Properties of emulsifier are shown in Table 1. Polymeric comb surfactant (DAPCO, Iran) was used as a binder for coating the microcapsules on fabric. Polyester fabric with plain weave was used as textile material. The amounts of material used in the experiments are shown in Table 2.

Micro- and nanoparticles based on the acrylic polymers as carriers have been prepared by various microencapsulation techniques including organic phase separation, solvent evaporation and spay drying technique; and in this research, microcapsules were prepared by solvent evaporation technique [4].

For microencapsulation process, rose essence and comb acrylic copolymer were dissolved in acetone separately and mixed together. At the next step, soft water and emulsifier was also mixed together, and when this solution was being mixed by a mixer at speed of 1200 RPM, the solution of copolymer and essence were calmly added to the water. This emulsion was mixed for one hour at 20-25°C; and at last, the temperature was raised to 70°C until microcapsules prepared completely. To make the microcapsules ready in narrow range size, the emulsion was filtered by paper filter. In the next step, %10 polymeric comb surfactant was blended with the filtered emulsion and sprayed on surface of polyester fabric. Curing process of coated fabric was carried out in oven for 15 min at 120°C. The assessment of microcapsules fastness against washing was evaluated according to ISO 105-C01.Finally the surface of polyester fabrics was studied using Scanning Electronic Microscope (SEM LEO 1455 VP). To measure the microcapsules size, Master size 2000, (Malvern Company) was used.

TABLE 1. Emulsion properties.

Acid Value (mgKOH/g)	Hydroxy Value (mgKOH/g)	HLB	M_w
18-20	88	12.3	650

RESULTS AND DISCUSSION

This research has been carried out for microencapsulation of rose essence with comb acrylic copolymer and its coating on polyester fabrics. Scanning electronic microscope was used for characterization of microcapsules. Figure 1 shows surface properties of polyester fibers after coating of microcapsules. Figure 1a shows that diameters of particles are about 400 nanometers and they spread levelness at the surface of fiber. The magnification of Figure 1a is 9000x. Figure 1b has shown similar results compared to Figure 1a and its magnification is 4000x. As can be seen in Figure 1c, particles have similar size and they are 400 nanometers. Surface properties of microcasule coated polyester fibers after two times standard washing process are shown in Figure 2. According to Figure 2, there was not considerable change in the number of particles at the fiber surface and microcasulated fibers show good wash fastness properties. It can be concluded that coating of microcapsules on polyester fibers with acrylic polymer is an effective process and fibers has good fastness against washing. Particle sizer was carried out for measuring of aggregated microcapsules in aqueous media. Particle evaluation showed that the size of some aggregated microcapsules is 40 μm.

TABLE 2. Amount of used material.

Sample	Copolymer (g)	Used Acetone For Copolymer (g)	Essence (g)	Used Acetone For Essence (g)	Soft Water (g)	Emulsifier %
1	10	50	2	15	100	1.5
2	10	50	2	15	300	1.5
3	10	50	2	15	400	1.5
4	5	25	2	15	100	1.5
5	15	75	2	15	100	1.5

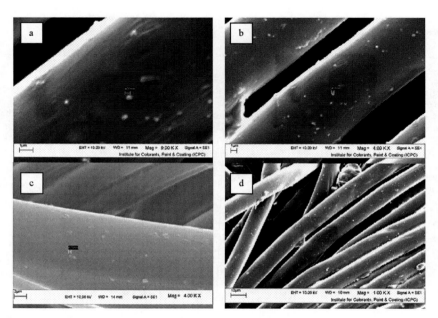

FIGURE 1. SEM images of coated micro- and nanocapsules on polyester fabrics before washing (1a: with magnification of 9000, 1b & 1c: with magnification of 4000x, 1d: with magnification of 1000x).

226

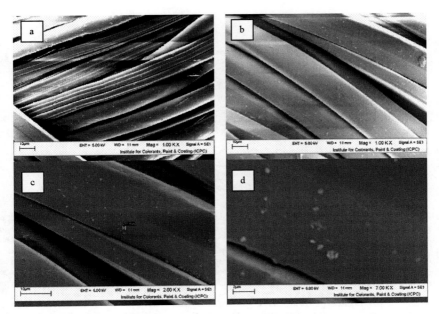

FIGURE 2. SEM images of coated micro- and nanocapsules on polyester fabrics after two wash cycles (1a: with magnification of 1000x, 1b: with magnification of 2000x, 1c & 1d: with magnification of 1000x).

CONCLUSION

Polyester fabrics were coated with fragrance microcapsules. Surface properties of fabric were evaluated with SEM. Particles have the size of 400 nm and they spread on the fiber surface. Microcapsules showed good properties against washing process.

REFERENCES

1. G. Nelson, "Application of Microencapsulation in Textiles", *International Journal of Pharmaceutics* **242**, 55 (2002).
2. F. Ulmann, "Microencapsulation", *Ulmann's Encyclopedia of Industrial Chemistry*, 6th Edition (John Wiley, 2003) pp. 733-746.
3. K. Othmer, "Microencapsulation", *Encyclopedia of Chemical Technology*, 3rd Edition, **15** (1981) pp. 470-491.
4. M. Athiowitz, "Nondegradable polymers for drug delivery", *Encyclopedia of controlled Drug Delivery* **2** (John Wiley, 1999). pp. 674-684.

Controlled Fabrication of Gelatin Nanoparticles as Drug Carriers

M. Jahanshahi[1], M. H. Sanati[2], Z. Minuchehr[2], S. Hajizadeh[1] and Z. Babaei[1]

[1] *Nanobiotechnology Research Laboratory, Faculty of Chemical Engineering, Noshirvani Mazandaran University of Technology, P. O. Box 484, Babol, Iran*
mmohse@yahoo.com
[2] *The National Research Center for Genetic Engineering and Biotechnology, Tehran, Iran*

Abstracts. In recent years, significant effort has been devoted to develop nanotechnology for drug delivery since it offers a suitable means of delivering small molecular weight drugs, as well as macromolecules such as proteins, peptides or genes by either localized or targeted delivery to the tissue of interest. Nanotechnology focuses on formulating therapeutic agents in biocompatible nanocomposites such as nanoparticles, nanocapsules, micellar systems, and conjugates. Protein nanoparticles (BSA, HAS and gelatin) generally vary in size from 50-300 nm and they hold certain advantages such as greater stability during storage, stability in vivo, non-toxicity, non-antigen and ease to scale up during manufacture over the other drug delivery systems. The primary structure of gelatin offers many possibilities for chemical modification and covalent drug attachment. Here nanoparticles of gelatin type A were prepared by a two-step desolvation method as a colloidal drug delivery system and the essential parameters in fabrication were considered. Gelatin was dissolved in 25 mL distilled water under room temperature range. Then acetone was added to the gelatin solution as a desolvating agent to precipitate the high molecular weight (HMW) gelatin. The supernatant was discarded and the HMW gelatin re-dissolved by adding 25 mL distilled water and stirring at 600 rpm. Acetone were added drop-wise to form nanoparticles. At the end of the process, glutaraldehyde solution was used for preparing nanoparticles as a cross-linking agent, and stirred for 12h at 600 rpm. For purification stage we use centrifuge with 600rpm for 3 times. The objective of the present study is consideration of some factors such as temperature, gelatin concentration, agitation speed and the amount of acetone and their effects on size and distribution of nanoparticles. Among the all conditions, 60° C, 50 mg/ml gelatin concentration, 75 ml acetone had the best result and the nanoparticle size was under 170 nm. The effect of these factors for synthesis of gelatine nanoparticle is strongly discussed.

Keywords: gelatin nanoparticles, fabrication, optimization, drug delivery.
PACS: 81.07.-b

INTRODUCTION

Nanoparticle formulation of anticancer drugs has become an important research area in cancer nanotechnology which can provide a way of sustained, controlled and targeted drug delivery to improve the therapeutic effects and reduce the side effects of the formulated drugs [1]. The main target of many pharmaceutical delivery systems is to deliver the drug directly to the specific cell types and is successful only when the drug through its delivery vehicle is internalised into cells [2]. Particulate colloidal carriers (i.e., liposomes or nanospheres or nanocapsules) were developed and are now proposed as a new approach for drug administration and vaccines [3]. These systems in general can be used to provide targeted (cellular / tissue) delivery of drugs, to improve oral bioavailability, to sustain drug/gene effect in target tissue, to solubilize drugs for intravascular delivery, and to improve the stability of therapeutic agents against enzymatic degradation (nucleases and proteases), especially of protein, peptide, and nucleic acids drugs [1]. The nanometer size-ranges of these delivery systems offer certain distinct advantages for drug delivery [4]. The sub-micron size of nanoparticles offers a number of distinct advantages over microparticles. Nanoparticles have in general relatively

CP929, *Nanotechnology and Its Applications, First Sharjah International Conference*
edited by Y. I. Salamin, N. M. Hamdan, H. Al-Awadhi, N. M. Jisrawi, and N. Tabet
© 2007 American Institute of Physics 978-0-7354-0439-7/07/$23.00

igher in tracellular uptake compared to microparticles. This was demonstrated in previous studies in which 100 nm size nanoparticles showed 2.5 fold greater uptakes compared to 1 mm and 6 fold higher uptakes compared to 10 mm microparticles [5]. The body distribution of colloidal drug delivery systems is mainly influenced by two physicochemical properties, particle size and surface characteristics [6].

Biodegradable nanoparticles can be synthesized from selected natural or synthetic macromolecules, such as serum albumin, polycyanoacrylates, polylactic-co-glycolic acid, and recently chitosan. Several researchers have investigated the use of gelatin as biomaterial to synthesize drug delivery systems. Gelatin nanoparticles have been used for delivery of different drugs, gene delivery, as carriers to deliver drug to lungs, and recently antibody modified gelatin nanoparticles were used to target lymphocytes, leukemic cells and primary T-lymphocytes [7]. Protein nanoparticles generally vary in size from 50-300 nm [8] and they hold certain advantages such as greater stability during storage, stability in vivo, non-toxicity, non-antigen [5] and ease to scale up during manufacture [6, 9] over the other drug delivery systems. Among the proteins, gelatin has a better chance to be approved as a biopolymer for nanoparticles since it has been used for decades as plasma expander. Several methods have been developed for the preparation of gelatin particles, such as emulsification, coacervation and desolvation. The emulsification method, in which an aqueous gelatin solution is emulsified in an oily phase and subsequently precipitated by cold or heat treatment, usually leads to the formation of microparticles [10, 11]. In order to obtain nanoparticles, desolvation methods are more appropriate 12].

In this work we studied the effect of different imperative factors on the fabrication of gelatin nanoparticle with desolvation method. Our goal was to consider these factors and optimize the particle size and size distribution of gelatin nanoparticles. The effected of manufacturing conditions such as temperature, gelatin concentration, agitation speed, acetone adding rate etc. upon the fabrication of such nanoparticle is strongly investigated. The next step of our study is using Taguchi design method to obtain the optimum condition of gelatin nanoparticles fabrication.

MATERIALS AND METHODS

Materials

All the chemicals were of reagent grade and were used without further purification. Gelatin type A (from porcine skin), glutaraldehyde, HCl and acetone were obtained from Sigma, Poole, UK. Double distilled water was used for all the experiments.

Fabrication and Purification of Gelatin Nanoparticles

The nanoparticles were prepared using a desolvation technique. 1.25 g gelatin was dissolved in 25 mL distilled water under constant heating temperature range. 25 mL acetone or ethanol was added to the gelatin solution as a desolvating agent to precipitate the high molecular weight (HMW) gelatin. The supernatant was discarded and the HMW gelatin re-dissolved by adding 25 mL distilled water and stirring at 600 rpm under constant heating. The pH of the gelatin solution was adjusted at 2.5. Acetone (75 mL) were added drop-wise to form nanoparticles. At the end of the process, 250 µL of 25% glutaraldehyde solution was used for preparing nanoparticles as a cross-linking agent, and stirred for 12h at 600 rpm. The particles were purified by threefold centrifugation and redispersion in 30% acetone in milliQ water. After the last redispersion, the acetone was evaporated using concentrator (speed vaccum).

Nanoparticles Characteristics and Size Distribution Determinations

For taking image with scanning electron microscopy (SEM), fifty microlitres of the nanoparticle preparation were freeze-dried (Emithech; model IK750, UK) on a polished aluminium surface. The particle size of the resulting nanoparticles was determined by photon correlation spectroscopy (PCS), Zetasizer 3000, Malvern Instruments, UK, with a He–Ne laser beam at a wavelength of 633 nm (scattering angle of 908).

RESULT AND DISCUSSION

Physical Characterization of Nanoparticles

A range of protein nanoparticle having broadly similar particles size and anionic characters to other nanoparticle such as adenovirus and plasmid DNA were fabricated based on simple coacervation. Average particle size in base of size distribution was calculated by PCS. In addition, the morphology and size distribution of the prepared nanoparticles from gelatin were examined with Atomic Force Microscopy and Scanning Electronic Microscope (figure 1). The shape of the nanoparticles demonstrated in SEM is spherical.

The Effect of Different Parameters on Nanoparticle Size

The effects of five parameters were examined in this research: temperature, rate of agitation, concentration of gelatin, concentration of acetone and cross-linker. The effect of these factors on particle size was illustrated in table 1, 2, 3, 4 and 5 respectively. In each experiment, all the conditions were kept constant and the only under research parameter changed.

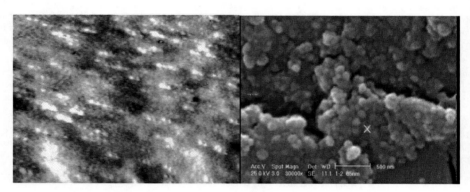

FIGURE 1. (left) The AFM image (right) the SEM image of Gelatin nanoparticles.

TABLE 1. The temperature effect on nanoparticle size.

Tempreture (^{0}C)	Size (nm)	Poly dispersity index
40	209.1	0.061±0.007
50	185.6	0.0202±0.012
60	169.7	0.0674±0.045

As it can be seem from the table 1, by increasing the temperature the particle size decreased. It was found that the preparation of nanoparticles at ambient temperature (25 °C) was not possible.

TABLE 2. Influence of the rate of agitation on nanoparticle size.

Rate of agitation(rpm)	Size (nm)	Poly dispersity index
500	258.7	0.0336±0.004
600	213.1	0.028±0.015
700	203.4	0.0595±0.065

230

Table 2 shows a minimum size of particle (203 nm) was obtained at agitation speed of 700 rpm. Generally, the size of particle is expected to reduce with increasing trend of agitation speed. The particle size was unaffected with agitation speed of higher than 700 rpm (data not shown). Tables 3 and 4 showed the effect of the concentration of gelatin and amount of acetone adding. In order to increase the concentration of gelatin and amount of acetone the nanoparticle size were decrease.

TABLE 3. The effect of the gelatin concentration on nanoparticle size.

Concentration of gelatin (mg/ml)	Size (nm)	Poly dispersity index
45	238	0.0402 ± 0.010
50	219.5	0.0160 ± 0.013
55	193	0.0237 ± 0.003

TABLE 4. Influence of the amount of acetone on nanoparticle size.

Acetone (ml)	Size (nm)	Poly dispersity index
70	303.5	0.0101 ±0.021
75	253.4	0.022 ± 0.011
80	211.2	0.009 ± 0.032

Table 5 showed the effect of concentration of glutaraldehyde on the size of gelatin nanopartcles. As it was shown in table 5, increasing the concentration of glutaraldehyde did not show any significant effect on the particle size of the nanoparticles.

TABLE 5. Influence of the cross-linker concentration on the particle size.

Acetone (ml)	Size (nm)	Poly dispersity index
250	208.0	0.0352 ±0.075
300	279.6	0.042 ± 0.021
350	212.4	0.013 ± 0.003

CONCLUSION

The protein nanoparticle as it has been assembled here, not only mimics the size and surface chemistry of nanoparticles such as viruses and plasmid, but also can be used as drug delivery vehicles in its own right. Gelatin type A was used in two step desolvation method for preparation of nanoparticles. The nanoparticles size fabricated here was influenced by several process variables including agitation speed, temperature and gelatin concentration and etc. the best result (minium size of the gelatin nanoparticles) were achieve in 60° C, 50 mg/ml gelatin concentration, 75 ml acetone and the nanoparticle was under 170 nm. Optimization of this fabrication method by Taguchi method for protein nanoparticles as drug delivery systems will be the subject of next publication.

ACKNOWLEDGMENT

The Authors would like to thank Noshirvani Mazandaran University of Technology for supporting this research. Also special thanks to The National Research Center for Genetic Engineering and Biotechnology, Iran.

REFERENCES

1. C. Coester, P. Nayyar, and J. Samuel, *European Journal of Pharmaceutic and Biopharmaceutics* **62**, 306–314 (2006).
2. A. Saxena, K. Sachin, H. B. Bohidar, and A. K. Verma. *Colloids and Surfaces B: Biointerfaces* **45**, 42-48 (2005).
3. P. Couvreur, R. Gref, K. Andrieux, and C. Malvy, *Progress in Solid State Chemistry* **34**, 231-235 (2006).
4. B L. Mua, P.H. Seowc, *Colloids and Surfaces B: Biointerfaces* **47**, 90–97 (2006).

5. S. K. Soppimath, T. M. Aminabhavi, A. R. Kulkarni, and W. E. Rudzinski *Journal of Controlled Release* **70**, 1-20 (2002).
6. S. S. Davis, L. Illum, E. T. Linson and S. S. Davis (Ed.), Site-specific Drug Delivery: Cell Biology, Medical and Pharmaceutical Aspects (John Wiley & Sons, Chichester, 1986).
7. M. Jahanshahi, H. Aghajani and TC Ling. *International Journal of Nanoscience and Nanotechnology* **1**, 9 (2005).
8. Sh. Azarmi, Y. Huang, H. Chend, S. McQuarriea, D. Abramse, W. Road, W. H. Finlayf, G. G. Millera, R. Löbenberga, *J. Pharm. Pharmaceut. Sci.* **9**, 124 (2006).
9. S. Balthasar, K. Michaelis, N. Dinauer, H.V. Briesen, J. Kreuter, K. Langer. *Biomaterials* **26**, 2723 (2005).
10. C. J. Coester, K. Langer, H. Van Briesen, and J. Kreuter, *J. Microencapsul.* **17**, 187 (2000).
11. C. Weber, C. Coester, J. Kreuter, and K. Langer, *Int J Pharm* **194**, 91 (2000).
12. J. Kreuter, *Journal of Controlled Release* **16**, 169 (1991).

MISCELLANEOUS

Composition Effect on the Optical and Electrical Parameters of Bi_xSe_{100-x} Thermal Evaporated Films

A. El-Korashy, M. Abdel Rahim, M. M. Hafiz and A. Z. Mahmoud

Physics Department, Faculty of Science, Assiut University, Assiut, Egypt
aelkorashy@yahoo.com

Abstract. A systematic study of the thermal, structural, optical and electrical properties of Bi_xSe_{100-x} glasses *(x = 5, 10, 15, 25 at %)* were carried out. The bulk materials were prepared by the usual melt quenching technique then thin films were deposited by thermal evaporation under 10^{-5} Torr vacuum. The DTA analysis indicated the dependence of the glass transition and crystallization temperatures on the compositions. The crystalline phases were identified using x-ray powder diffraction and scanning electron microscope techniques. The mechanism of the optical absorption followed the rule of direct transition. The optical energy gap decreases and width of localized states increases with increasing Bi content. The dependence of the electrical conductivity on the composition has been investigated. The $\sigma - T$ dependence for all considered compositions shows semiconductor materials behavior. The conduction phenomena of the investigated thin films proceeded through two distinct mechanisms. In the high temperature region the thermally activated conduction may be represented through the extended states. In the low temperature region (less thermally activated) can be represented by the hopping conduction through the localized states.

Keywords: Optical, electrical, thin films, Bismuth chalcogenides.
PACS: Category 60, 70.

INTRODUCTION

Amorphous chalcogenide glasses are of great interest due to their importance in preparing optical recording films [1,2]. Films are switched between amorphous and crystalline states using the heat of a focused Laser beam. Large reflectivity differences between amorphous and crystalline states are then used to store and retrieve information. Many amorphous semiconducting glasses, in particular selenium, exhibit a unique property of the reversible transformation [3]. The selection of Se is because of its wide commercial applications in photocells, switching, and memory devices. The amorphous Selenium (a-Se) pure state has many disadvantages because of its short lifetime and low sensitivity [4], but the added impurities have a considerable effect on its physical properties [5,6].

Most of selenium doped with other elements exhibits p-type conduction and the electrical conductivity is slightly affected by doping. This is due to the presence of charged defects which pin the Fermi level near the middle of the band gap [7]. In particular, Bismuth-doped Selenium exhibits n-type conduction [8]. Bi_xSe_{100-x} systems are characterized by the low crystallization and melting temperatures as the addition of Bismuth facilitates the crystallization process [4, 9]. According to Fleury et al. [10] above 3 at % of Bi, the Bi_xSe_{100-x} samples were no longer amorphous. Muñoz et al. [4] obtained amorphous films up to 27 at % of Bi. Finally, Y. Sripathi [11] prepared amorphous films containing 50 at % of Bi. Optical memory effects in amorphous semiconductor films have been investigated for various applications [1, 12]. The present work reports the effect of increasing Bismuth content on the optical and electrical properties of Bi_xSe_{100-x} thin films.

EXPERIMENTAL

The bulk chalcogenide glasses of the Bi_xSe_{100-x} system were prepared by a melt quenching technique. The Bi and Se samples (99.99% pure) were weighed, mixed and sealed into a clean Silica tube under vacuum of 10^{-5} Torr. The tubes were heated stepwise to 1000 °C and kept at the maximum temperature for 24 hours with frequent shaking

CP929, *Nanotechnology and Its Applications, First Sharjah International Conference*
edited by Y. I. Salamin, N. M. Hamdan, H. Al-Awadhi, N. M. Jisrawi, and N. Tabet
© 2007 American Institute of Physics 978-0-7354-0439-7/07/$23.00

to ensure alloy homogeneity, and then were quenched into ice water. Thin films were prepared by the thermal evaporation method using a Molybdenum boat. Deposition was done onto ultrasonically cleaned glass substrates at pressure of 10^{-5} Torr using a standard (Edward E306) coating unit. The glass substrates were kept at room temperature during the evaporation process. The evaporation rates as well as the thickness (about 150 nm) were controlled using a quartz crystal monitor (Edward FTM5). A constant evaporation rate (2.5 nm s^{-1}) was used to deposit the required films. A Philips model (PW1710) X-ray diffractometer, using Cu radiation (λ=1.542 Å) operated at 40 KV - 30 mA with a scanning speed of $3.6°min^{-1}$, was used to investigate the film structure. The Optical transmittance (T) and reflectance (R) spectra of the investigated samples were obtained at room temperature in the wavelength range (300 - 900 nm) using a double beam scanning spectrometer (Shimadzu 160A). The electrical measurements were carried out on films with evaporated gold electrodes using a conventional circuit involved a (Keithly 610C) electrometer.

RESULTS AND DISCUSSION

Structure Results

A complete set of DTA thermograms for Bi_xSe_{100-x} glasses was measured at heating rate 10 K/min as shown in Fig. 1.

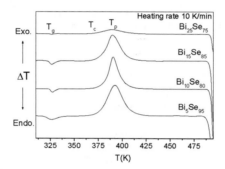

FIGURE 1. Typical DTA Traces at heating rate 10 K/min for the four compositions.

The DTA curves of all compositions show a single glass transition and single stage crystallization peak was observed. The values of crystallization temperatures T_c and the peak temperatures T_p are shifted towards the lower temperature as Bi content increased. The glass temperature T_g increased with increasing Bi content. On the other hand, the morphology of the samples annealed at 383 K (near T_c) was examined using a scanning electron microscope (SEM). Fig. 2 shows the scanning micrograph of specimens of compositions Bi_5Se_{95} and $Bi_{25}Se_{75}$ (highest and lowest Bi content) annealed at 383 K for 1 hour.

Fig. 2-a shows the existence of polycrystalline structure consisting of crystallites embedded in an amorphous matrix while Fig. 2-b shows the surface microstructure which consists of different crystalline phases and some of the crystallized particles which are interconnected. In general, crystallites occupy most of the structure. Such observation indicates that the crystalline morphology increases in size with increasing Bi content.

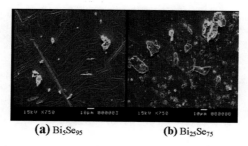

(a) Bi₅Se₉₅   (b) Bi₂₅Se₇₅

FIGURE 2. SEM photograph for the compositions annealed at 383 K for 1 hour.

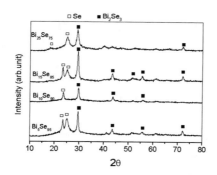

FIGURE 3. X-ray diffraction pattern for the compositions annealed at 383 K for 1 hour.

To identify the crystalline phases, X-ray diffraction was examined for these annealed samples. Fig. 3 shows the diffraction pattern for these compositions. The analysis reveals that the crystalline peaks are due to Bi_2Se_3 and Se phases.

Optical Results

The absorption coefficient (α) was calculated from the following expression [13]

$$T = \frac{(1-R)^2 \exp(-\alpha d)}{1 - R^2 \exp(-2\alpha d)}, \tag{1}$$

where d (cm) is the film thickness. This equation neglects the substrate effects and it can be applied only for homogeneous thin films.

According to Tauc [14] there are three regions in the optical absorption spectra, the high absorption region ($\alpha \geq 10^4$ cm^{-1}), the exponential region ($1 \leq \alpha \leq 10^4$ cm^{-1}) and the weak absorption tail ($1 \leq \alpha$ cm^{-1}) which originates from the defects and impurities.

Fig. 4 shows the relation between Log (α) and the photon energy (hυ) for the prepared films (150 nm in thickness) in the high-absorption region, and shows two absorption regions in all samples. The first region is in the range ($1.8 <$ h$\upsilon < 2.7$ eV) and the second is in the range ($2.7 <$ h$\upsilon < 3.4$ eV). Tauc [14] and Davis and Mott [15] derived an expression relating the absorption coefficient (α) to the photon energy (hυ)

$$\alpha = \frac{A(\alpha h\upsilon - E_{opt})^r}{h\upsilon}, \tag{2}$$

where A is a constant ($= 4\pi\sigma/ncE_c$), c is the speed of light and σ is the extrapolated DC conductivity at ($T = \infty$). E_c is a measure of the extent of band tailing, n is the refractive index and E_{opt} is the optical absorption energy gap, ($r = 1/2$)

237

for the direct allowed transition. Fig. 5-a shows the plots of $(\alpha h\upsilon)^2$ vs. $h\upsilon$ for the prepared thin films in the range (2.7 > $h\upsilon$ > 1.8 eV). The results show that E_{gd1} decreases from 2.16 to 2.01 eV with increasing Bi content. Fig. 5- shows the plots of $(\alpha h\upsilon)^2$ vs. $h\upsilon$ in the range of (3.4 > $h\upsilon$ > 2.7 eV). E_{gd2} decreases from 2.79 to 2.53 eV with increasing Bi content.

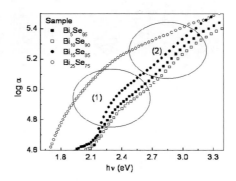

FIGURE 4. Dependence of Log (α) on $h\upsilon$ for the as-deposited Bi_xSe_{100-x} thin films.

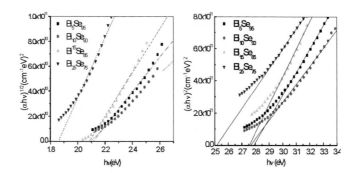

FIGURE 5. (a) $(\alpha h\upsilon)^2$ against ($h\upsilon$) for Bi_xSe_{100-x} films in the first region (E_{g1}).(b) $(\alpha h\upsilon)^2$ against ($h\upsilon$) for Bi_xSe_{100-x} films in the second region (E_{g2}).

In the exponential region, we use the Urbach relation [19] to compute the width of the localized states at the band gap tails (E_e)

$$\alpha = \alpha_0 e^{h\upsilon / E_e}, \tag{3}$$

where α_0 is a constant.

TABLE 1. The activation energy, DC conductivity, the degree of disorder, the density of localized states at the Fermi level, the hopping distance and hopping energy for the as-deposited Bi_xSe_{100-x} films.

Bi %	E_{g1}	E_{g2}	E_e
5	2.16	2.79	0.022
10	2.14	2.77	0.025
15	2.12	2.76	0.027
25	2.01	2.53	0.056

As shown in Fig. 6 plotting Ln (α) vs. (hυ) we get the values of the band tail width (E_e) of the localized states from the slope of a straight line. The calculated values of E_e (0.022-0.056 eV) increase with increasing Bi content. The obtained values of E_{gd1}, E_{gd2} and E_e are listed in Table 1. Now we notice that the values of E_{gd2} are in general higher than those of E_{gd1}. The energy gap slowly decreases as Bi content decreases from 5 to 15 at % but for Bi > 15 at % one can observe a large change. E_e exhibits an opposite behavior to that of E_{gd1} and E_{gd2}. Therefore, increasing the concentration of Bi atoms could decrease the glass bond energies and consequently its optical energy gap [17, 20]. Bi_xSe_{100-x} thin films show an unusual change in the shape of the $(\alpha h\upsilon)^2$ - (hυ) curve after the primary transition, the tail appears. The two transitions observed with two energy gaps may be attributed to the presence of a secondary phase [13].

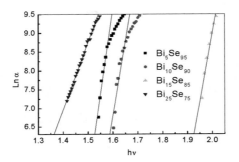

FIGURE 6. Dependence of the absorption coefficient on the photon energy for as-deposited Bi_xSe_{100-x} thin films.

Electrical Conductivity Results

The effect of Bi on the electrical conductivity (σ) of the deposited Bi_xSe_{100-x} films of thickness 3000 Å deposited onto glass substrates held at room temperature was studied. The resistance was measured during heating the sample in the temperature range (300–400 K).

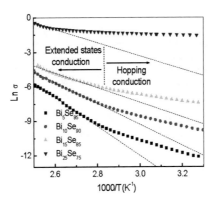

FIGURE 7. Ln (σ) vs. (1000/T).

Fig. 7 shows Ln (σ) vs. (1000/T). Each curve has two regions: the hopping conduction region and the regular band conduction in extended states at (T>T$_g$). So, the measured conductivity is the sum of two components [21] and will be written as

$$\sigma = \sigma_{hop} - \sigma_{ext},\qquad(4)$$

where σ_{hop} is the contribution of conduction due to hopping between the localized states and σ_{ext} is the conduction between the extended states. In the low temperature region (300K $\leq$ T $\leq$ 350K), the conduction occurs via variable-range hopping of the charge carriers in the localized states near the Fermi level, and is characterized by Mott's relation [21, 24].

$$\sigma = \frac{\sigma_0}{\sqrt{T}} \exp\left(-\frac{T_0}{T}\right)^{1/4},\qquad(5)$$

where

$$T_0 = \frac{18\alpha^3\lambda}{k_B N(E_f)},\qquad(6)$$

in which where T$_0$ is the degree of disorder, N(E$_f$) is the density of localized states at the Fermi level, k$_B$ the Boltzman constant, λ is a dimensionless constant and α is the coefficient of the exponential decay of the localized state wave function (i.e.α^{-1} represents the degree of localization), and it is assumed to be 0.125 Å^{-1} [25].

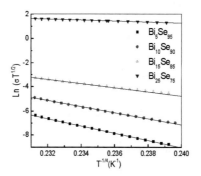

FIGURE 8. Ln (σ T$^{1/2}$) vs. T$^{-1/4}$.

Fig. 8 shows the linear plots of Ln (σ_{hop}T$^{1/2}$) vs. (1/T)$^{1/4}$ for the investigated compositions. From the slopes of the resulting straight lines, the T$_0$ values can be calculated and listed in Table 2. Also the density of localized states at the Fermi levels N (E$_f$) has been calculated.

Fig. 9 shows the increase of N (E$_f$) and the decrease of the average hopping energy W with increasing the Bi content. These results are in agreement with Majeed Khan et al. [26].

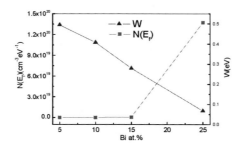

FIGURE 9. The density of localized states at the Fermi level N(E$_f$) and the average hopping energy W vs. Bi content.

According to Mott [27] and Davis et al. [15] two other hopping parameters can be calculated, the hopping distance R (cm) and hopping energy W (eV) which are given by

$$R = \left[\frac{9}{8\pi K_B T N(E_f)} \right]^{1/4},$$ (7)

$$W = \frac{3}{4\pi R^3 N(F_f)}.$$ (8)

The calculated values of R and W for the investigated films are listed in Table 2. It is found that W and T_o decrease with increasing Bi concentration, since T_o represents the degree of disorder and α^{-1} is the degree of localization, it follows that the amorphicity of the samples decreases with increasing content.

The increase of Bi leads to the formation of negatively charged Bi^- and three coordinated positively Se^+ [28]. The pinning of Fermi energy in the middle of the band gap is a result of the equilibrium between negatively and positively charged dangling bonds, D^- and D^+ [28]. The introduction of Bi^- perturbs the equilibrium, leading to the unpinning of the Fermi energy associated with increasing the density of localized states [26].

On the other hand, increasing Bi content increases the Bi-Se bonds, and the bonds between the chalcogenide atoms (e.g., Se-Se bond) decrease. This process is associated with the increase in the value of the density of localized states.

As shown in Fig. 6, in the high temperature region, the relation between Ln (σ) against (1000/T) can be described by an Arrhenius relation

$$\sigma = \sigma_o \exp(-\Delta E / k_B T),$$ (9)

where ΔE is the activation energy for DC conduction, k_B is the Boltzmann constant and σ_0 is the pre-exponential factor. σ and ΔE are calculated at T = 307.35 K are shown in Fig. 10. ΔE decreases and σ increases with increasing Bi content. This confirms that the changes in σ are connected to the corresponding change in ΔE.

The calculated values of ΔE are found to be about half the magnitude of the optical energy gap. This means that E_f is not far from the center of the mobility gap.

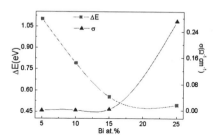

FIGURE 10. Dependence of the activation energy and DC conductivity on the Bi content in the investigated films.

This variation of DC conductivity and activation energy was also observed on SeGeBi [29] and on SeGeTeBi [30] alloys for 9 at. % of Bi. The increase in DC conductivity and the corresponding decrease in activation energy is found to be associated with the shift of Fermi level in impurity doped chalcogenide glasses [31, 32]. This behavior is essentially due to the metallic character of bismuth. Bi acts as a metal added in a chalcogenide matrix, which leads to the unpinning of the Fermi level; then the Fermi level shifts towards the conduction band and the conduction becomes of n type [33]. However, it is also pointed out that the increase in conductivity could be caused by the increase in hopping conduction through defect states associated with the impurity atoms [32]. The conduction takes place either in the extended states above the mobility edge or hopping in the localized states.

REFERENCES

1. L. K. Malhotra et al., Bull. Mater. Sci, **18** (1995) 725.
2. K. A. Rubin and M. Chen, Thin Solid Films **181** (1989) 129.

3.   K. Tanaka, Phys. Rev. B **39** (1989) 1270.
4.   A. Muñoz, F.L. Cumbrera and R. Márquez, Thin Solid Films **186** (1990) 37.
5.   J. Vázquez , C. Wagner, P. Villares, R. Jiménez-Garay, J. Non-Cryst. Solids **235** (1998) 548.
6.   M. F. Kotkata, F. M. Ayad and M. K. El-Mously, J. Non-Cryst. Solids, 33 (1979) 13.
7.   C. Vautier, J.M. Saiter and T. Derrey, J. Non-Cryst. Solids **103** (1988) 65.
8.   N. Tohge, T. Minami, Y. Yamamoto, and M. Tanaka, J. Appl. Phys. **51** (1980) 1048.
9.   M.M. Hafiz, O. El-Shazly, and N. Kinawy, Appl. Surf. Sci. **171** (2001) 231.
10. G. Fleury, A. Hamou, C. Lhermitte, and C. Viger, Phys. Stat. Solidi (a) **83** (1984) K103.
11. Y. Sripathi, G.B. Reddy, and L.K. Malhotra, J. Mater. Sci., Mater. Electron. **3** (1992) 164.
12. K.A. Rubin, and M. Chen, Thin Solid Films **181** (1989) 129.
13. J.I.Pankove, Optical Processes in Semiconductors (Dover, New York, 1971) p.103.
14. J.Tauc, in: F.Abeles (Ed.) Optical Properties of Solids (North Holland, Amsterdam, 1970) p. 903.
15. E.A.Davis, N.F.Mott, Phil. Mag. **22** (1970) 903.
16. A. El-Korashy, M. A. Abdel-Rahim, H. El-Zahed, Thin Solid Films **338** (1999) 207.
17. D.J. Sarrach, J.P. de Neufville, J. Non-Cryst. Solids **22** (1976) 245.
18. A.El-Korashy, N.El-Kabany and H.El-Zahed, Physica B **365** (2005) 55.
19. F. Urbach, Phys. Rev. **92** (1953) 1324.
20. M. A. Majeed Khan, M. Zulfequar, M. Husain, Optical Materials **22** (2003) 21.
21. N. F. Mott, Philos. Mag. **19** (1969) 835.
22. A. K. Jonscher, J. Vac. Sci. Technol. **8** (1971) 135.
23. R. D. Gould, B. B. Ismail, Phys. Stat. Sol. (a) **134** (1992) K65.
24. H. Mori, K. Cotoch, H. Sakata, J. Non-Cryst. Solids **183** (1995) 22.
25. S. Mahadevan, A. Giridhar, K.J. Rao, J. Phys. C **10** (1977) 4499.
26. M. A. Majeed Khan, M. Zulfequar, M. Husain, Physica **B 322** (2002) 1.
27. S. Okano, M. Suzuki, K. Imura, N. Fukada, A. Hiraki, J. Non-Cryst. Solids **59–60** (1983) 969.
28. N. F. Mott and E.A. Davis, Electronic Processes in Non-Crystalline materials (1978).
29. N. Tohge, T. Minami, M. Tanaka, J. Non. Cryst. Solids 37 (1980) 23.
30. P. Negals, M. Retti, S. Vikrov, J. de Phys. **42** (C 4) (1981) 907.
31. B. T. Kolomiets, E. A. Lebedev, N. A. Rogachev, Fiz. Tekh. Popuorov. **8** (1974) 545.
32. H. Fritsche, P. J. Gaczi and M. A. Kastner, Phil. Mag. B **37** (1978) 593.
33. E. Ebenzer, K.R. Murali, Mary Juliana Chockalingam, V.K. Venkatesan, J. of Mat. Sci. **23** (1988) 386.

Recent Results on Topological Indices of Nanotubes

B. Taeri and M. Eliasi

Department of Mathematical Sciences, Isfahan University of Technology, Isfahan, Iran
b.taeri@cc.iut.ac.ir

Abstract. Topological indices of nanotubes are numerical descriptors that are derived from graph of chemical compounds. Such indices based on the distances in graph are widely used for establishing relationships between the structure of nanotubes and their physico-chemical properties. Harold Wiener in 1947 introduced the notion of path number of a graph as the sum of the distances between two carbon atoms in the molecules, in terms of carbon-carbon bound. The Wiener index of graph G is defined as $W(G)=1/2 \sum_{u,v \in V(G)} d(u,v)$, where $V(G)$ is the set of vertices of the graph and $d(u,v)$ is the distance between two vertices u,v. The hyper Wiener index of G is defined by $WW(G)=1/2\ W(G) +1/4\ \sum_{u,v \in V(G)} d(u,v)^2$. In this paper we present some new results on topological indices of nanotubes and calculate hyper Wiener index of some nanotubes.

Keywords: Topological indices, Nanotubes, Nanotorus.
PACS: 61.46.Fg

INTRODUCTION

In 1991 Iijima[1] discovered Carbon nanotubes as multi walled structures. Carbon nanotubes show remarkable mechanical properties. Experimental studies have shown that they belong to the stiffest and elastic known materials. These mechanical characteristics clearly predestinate nanotubes for advanced composites.Carbon nanotubes are cylindrical structures of nanometric size, based on a hexagonal lattice of carbon atoms (see Figure 1).

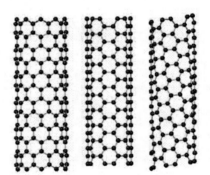

FIGURE 1. Three types of single-walled carbon nanotubes (C_6)   Zigzag,   Armchair,   Chiral .

These structures can be thought of as rolled two-dimensional graphene sheets. Their dimensions are typically a few nanometers across and up to 100 micrometers long. Since the beginning of the last decade carbon nanotubes have

CP929, *Nanotechnology and Its Applications, First Sharjah International Conference*
edited by Y. I. Salamin, N. M. Hamdan, H. Al-Awadhi, N. M. Jisrawi, and N. Tabet
© 2007 American Institute of Physics 978-0-7354-0439-7/07/$23.00

received remarkable attention form the research communities as it has been found that they poses unique electronic properties.

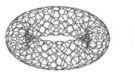

FIGURE 2. A C$_6$ nanotorus(Zigzag):Side view; Top view.

This makes nanotubes suitable for applications in micro electronics as they might be used to construct efficient diodes, transistors and displays. One of these properties is that nanotubes can behave as quantum wires. Carbon nanotubes show remarkable mechanical properties. Experimental studies have shown that they belong to the stiffest and elastic known materials. These mechanical characteristics clearly predestinate nanotubes for advanced composites. There are also various kinds of closed nanotubes, called nanotorus (see Figure 2) and also other type of nanotubes and nanotourus (see Figure 3).

TUC$_4$C$_8$(S) [28,80]   TUC$_4$C$_8$(R) [24,64]

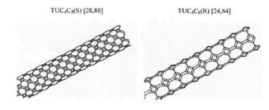

FIGURE 3. Nanotubes covered by **C$_4$C$_8$**.

TOPOLOGICAL INDICES

A topological index is a real number related to a molecular graph. It must be a structural invariant, i.e., it does not depend on the labelling or the pictorial representation of a graph. With hundreds of topological indices one would expect that most molecules could be well characterized and their physicochemical properties correlated with the available descriptors[2]. Topological indices of nanotubes are numerical descriptors that are derived from graph of chemical compounds. Such indices based on the distances in graph are widely used for establishing relationships between the structure of nanotubes and their physicochemical properties. Topological indices are a convenient method of translating chemical constitution into numerical values that can be used for correlations with physical properties studies. This method has been introduced by Harold Wiener as a descriptor for explaining the boiling points of paraffins[3]. The Wiener index is oldest topological indices. In 1947 chemist Harold Wiener developed the most widely known topological descriptor, the Wiener index, and used in to determine physical properties of types of alkanes known as paraffins. He introduced the notion of path number of a graph as the sum of the distances between two carbon atoms in the molecules, in terms of carbon-carbon bound. Since then, the spectrum of physico-chemical and biological properties enlarged continuously and several analogues have been defined. Numerous of its chemical applications were reported and its mathematical properties are well understood.

In a chemical language, the Wiener index is equal to the sum of all shortest carbon-carbon bond paths in a molecule. In a graph theoretical language, the Wiener index is equal to the count of all shortest distances in a graph. The Wiener index of a molecular graph provides a rough measure of the compactness of the underlying molecule. It was demonstrated[17] that the Wiener index is related to the molecular surface area. As a result, the Wiener index is reasonably well-correlated with many physical and chemical properties of organic compounds, and chemists are hence interested in computing it for a variety of classes of graphs. For further details on the mathematical properties and chemical applications of the Wiener index see Refs. 4, 5, the reviews and the references cited therein.

SOME ALGEBRAIC DEFINITION

We now recall some algebraic definitions that will be used in the paper. Let G be an undirected connected graph without loops or multiple edges. The set of vertices and edges of G are denoted by V(G) and E(G) respectively. If e is an edge of G connecting the vertices u and v of G, then we write e=uv. The distance between a pair of vertices i and j of G is denoted by d(u,v). The degree of a vertex $u \in V(G)$ is the number of vertices joining to u and denoted by deg(u). The (u,v) entry of the adjacency matrix of G is denoted by A(u,v). The Wiener index of the graph G is the half sum of distances over all its vertex pairs (u,v)

$$W(G) = \frac{1}{2} \sum_{u,v \in V(G)} d(u,v). \tag{1}$$

The distance of a vertex u of G is defined as

$$d(u) = \frac{1}{2} \sum_{x \in V(G)} d(u,x) \tag{2}$$

So we have

$$W(G) = \frac{1}{2} \sum_{u,v \in V(G)} d(u,v) = \frac{1}{2} \sum_{u \in V(G)} d(u). \tag{3}$$

Ivan Gutman[6] has introduced a generalization of the Wiener index for cyclic graphs called Szeged index. The main advantage of the Szeged index is that it is a modification of Wiener index for cyclic graphs; otherwise, it coincides with Wiener index. The Szeged index is obtained as a bond additive quantity where bond contributions are given as the product of the number of atoms closer to each of the two end points of each bond. Let u and v be two adjacent vertices of the graph G and e=uv be the edge between them. Let $B_u(e)$ be the set of all vertices of G lying closer to u than to v and $B_v(e)$ be the set of all vertices of G lying closer to v than to u, that is

$$B_u(e) = \{x \mid x \in V(G), \ d_G(x,u) < d_G(x,v)\} \tag{4}$$

$$B_v(e) = \{x \mid x \in V(G), \ d_G(x,v) < d_G(x,u)\}. \tag{5}$$

Let $n_u(e) = |B_u(e)|$ and $n_v(e) = |B_v(e)|$. The Szeged index of G is defined as

$$Sz(G) = \sum_{e \in E(G)} n_u(e)n_v(e). \tag{6}$$

The molecular topological index or Schultz index has been defined by Schultz[7]:

$$MTI(G) = \frac{1}{2} \sum_{\{u,v\} \subseteq V(G)} \deg(u)(d(u,v) + A(u,v)). \tag{7}$$

The Balaban index[8] of G calculates the average distance sum connectivity index according to the equation

$$J(G) = \frac{m}{\mu+1} \sum_{uv \in E(G)} [d(u)d(v)]^{-0.5}, \tag{8}$$

245

where m is the number of edges in G and μ is the cyclomatic number of G. The cyclomatic number μ of a connected graph G is defined as $\mu(G)=|E(G)|-|V(G)|+1$. The Balaban index calculates the average distance sum connectivity index and measures the ramification and it tends to increase with molecular ramification. It has been satisfactory correlated with octane numbers of alkanes.

Khadikar[9] defined a new topological index and named it Padmakar-Ivan index. They abbreviated this new topological index as PI. This newly proposed topological index does not coincide with the Wiener index for acyclic molecules. It is defined as $PI(G) = \sum_{e \in G}[n_{eu}(e|G)+ n_{ev}(e|G)]$, where $n_{eu}(e|G)$ is the number of edges of G lying closer to u than to v and $n_{ev}(e|G)$ is the number of edges of G lying closer to v than to u.

For some new results on topological indices of nanotubes we encourage to consult Refs. 10-44.

COMPUTING HYPER WIENER OF ZIGZAG POLYHEX NANOTORUS

In this section we derive an exact formula for the hyper Wiener index of $G:=HC_6[p,q]$, the zigzag polyhex nanotorus. For a vertex $u \in V(G)$ we define

$$d'(u) = \sum_{v \in V(G)} d(u, v)^2, \quad dd(u) = \sum_{v \in V(G)} [d(u) + d'(u)]. \tag{9}$$

Then

$$WW(G) = \frac{1}{4} \sum_{v \in V(G)} d(u) + \frac{1}{4} \sum_{v \in V(G)} d'(u) = \frac{1}{4} \sum_{v \in V(G)} dd(u). \tag{10}$$

Now for vertices u and v we define the hyper distance between u and v, as $dd(u,v)=d(u,v)+d(u,v)^2$. In the following lemma we give a formula for the hyper distances of one white (black) vertex of level 0 of the graph G to all vertices on the level $k < q/2$.

Lemma 1 In the graph G the summation, $\sum_{x \in \text{level } k} dd(x_{0,2i}, x)$ of hyper distances of one white vertex of level 0 to all vertices on level k (where $k<q/2$) is given by

$$\begin{cases} \frac{1}{6}p + \frac{5}{3}k + 5k^2 + kp + k^2p + \\ \qquad \frac{1}{4}p^2 + \frac{1}{2}kp^2 + \frac{1}{12}p^3 + \frac{10}{3}k^3 & \text{if } 0 \le k < \frac{p}{2} \\ p(2k+1)^2 & \text{if } \frac{p}{2} \le k \end{cases} \tag{11}$$

and the summation $\sum_{x \in \text{level } k} dd(x_{0,2i-1}, x)$ of hyper distances of one black vertex of level 0 to all vertices on level k (where $k<q/2$) is given by

$$\begin{cases} \frac{1}{6}p - \frac{1}{3}k - 3k^2 + kp + k^2p + \frac{1}{4}p^2 \\ \qquad + \frac{1}{2}kp^2 + \frac{1}{12}p^3 + \frac{10}{3}k^3 & \text{if } 0 \le k < \frac{p}{2} \\ 4k^2p & \text{if } \frac{p}{2} \le k \end{cases} \tag{12}$$

for all $i=1,2,\ldots, p$. We call these summation ww_k and bb_k respectively.

Proof: See Lemma 1 of Ref. [33].

Corollary 2 For each $u \in V(G)$ we have $D:=dd(u)=bb_0+bb_1+\ldots +bb_{q/2}+ww_1+\ldots +ww_{q/2-1}$.

Proof: At first note that the lattice is symmetric (with respect to the leve k). So it is suffices to consider x_{01} and x_{02}. For other black (white) vertices the argument is similar. Now we begin with x_{01}. Let $B_1 = \{k|\ 0 \leq k < q/2\}$ and $B_2 = \{k|\ q/2 < k \leq q-1\}$. We have

$$dd(x_{01}) = \sum_{v \in V(G)} dd(x_{01,,}, v)$$

$$= \sum_{v \in B_1} dd(x_{01,,}, v) + bb_{q/2} + \sum_{v \in B_2} dd(x_{01}, v). \qquad (13)$$

But,

$$\sum_{v \in B_1} dd(x_{01}, v) = \sum_{v \in \text{level } 0} dd(x_{01}, v) + \sum_{v \in \text{level } 1} dd(x_{01}, v) + \cdots$$

$$+ \sum_{v \in \text{level } q/2-1} dd(x_{01}, v) \qquad (14)$$

$$= bb_0 + bb_1 + \cdots + bb_{q/2-1}. \qquad (15)$$

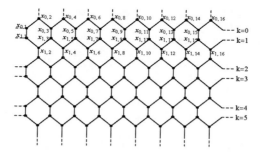

FIGURE 4. The lattice of a zigzag nanotorus with p=8, q=6.

For computing the last sum we consider the tori that can be built up from two halves collapsing at level 0. The top part is formed of the lines of B_2 that x_{01} is such as a black vertex. So by a changing index and using the proof of the Lemma 1, we can obtain that

$$\sum_{v \in B_2} dd(x_{01}, v) = \sum_{v \in \text{level } q-1} dd(x_{01}, v) + \sum_{v \in \text{level } q-2} dd(x_{01}, v) + \cdots$$

$$+ \sum_{v \in \text{level } q/2+1} dd(x_{01}, v) \qquad (16)$$

$$= ww_1 + ww_2 + \cdots + ww_{q/2-1}.$$

which completes the proof.

Theorem The hyper Wiener index, WW(G), of $G := HC_6[p,q]$ nanotori is given by

$$\begin{cases} \dfrac{1}{192} pq^2[-16 + 16p - 20q + 4p^3 + 6p^2q + 4pq^2 \\ \qquad + 12p^2 + 5q^3 + 4q^2 + 12pq] \text{ if } q \leq p, \\ \dfrac{1}{192} pq^2[3p^3 + 4p^2 - 12p - 6 + 16q^3 + 24q^2 + 8q] \text{ if } p < q. \end{cases} \qquad (17)$$

Proof: We have

$WW(G) = \frac{1}{4} \sum_{u \in V(G)} dd(u) = \frac{1}{4} \sum_{u \in V(G)} D = \frac{1}{4} D \mid V(G) \mid = \frac{1}{4} Dpq.$ First suppose that q $\leq$p-2. In this case q/2< p/2, so by Corollary 2 and Lemma 1 we can obtain that

$$D = bb_0 + bb_1 + \cdots + bb_{q/2} + ww_1 + \cdots + ww_{q/2-1}$$

$$= \frac{q}{48}(-16 + 16p - 20q + 4p^3 + 6p^2q + 4pq^2 +$$

$$12p^2 + 5q^3 + 4q^2 + 12pq). \tag{18}$$

Hence, in this case

$$WW(G) = \frac{1}{192} pq^2 (-16 + 16p - 20q + 4p^3 +$$

$$6p^2q + 4pq^2 + 12p^2 + 5q^3 + 4q^2 + 12pq). \tag{19}$$

Now suppose that p$\leq$q-2. In this case, p/2-1<q/2-1, so by Corollary 2 and Lemma 1 we can obtain that

$$D = bb_0 + bb_1 + \cdots + bb_{q/2} + ww_1 + \cdots + ww_{q/2-1}$$

$$= \frac{p}{48} p[3p^3 + 4p^2 - 12p - 16 + 16q^3 + 24q^2 + 8q]. \tag{20}$$

Hence

$$WW(G) = \frac{pq}{4} \frac{p}{48} p(3p^3 + 4p^2 - 12p - 16 + 16q^3 + 24q^2 + 8q)$$

$$= \frac{1}{192} p^2q(3p^3 + 4p^2 - 12p - 16 + 16q^3 + 24q^2 + 8q). \tag{21}$$

Similarly, we can handle the case p=q.

CONCLUSION

We reviewed some topological indices of molecular graphs and presented some recent references on topological indices of nanotubes and nanotori. Also we have introduced a method for finding the Wiener and Hyper Wiener indices of nanotubes and computed the Hyper Wiener index of zigzag polyhex nanotorus.

REFERENCES

1. S. Iijima, *Nature* **354**, 56-58 (1991).
2. R. Todeschini and V. Consonni, Handbook of Molecular Descriptors (Wiley, Weinheim, 2000).
3. H. Wiener, *J. Am. Chem. Soc.* **69**, 17-20 (1947).
4. I. Gutman, S.Klavžar and B.Mohar (eds.), *MATCH Commun. Math. Comput. Chem.***35**, 1–259 (1997).
5. I. Gutman, S.Klavžar and B.Mohar (eds.), *Discrete Appl. Math* **80**, 1–113 (1997).
6. I. Gutman, *Graph Theory Notes New York* **27**, 9-15 (1994).
7. H.P.Schultz, *J.Chem. Inf. Comput Sci.* **29**, (1989) 227-228.
8. A. Balaban, *Chem. Phys. Lett.* **89**, 399-404 (1982).
9. P. V. Khadikar, *Nat. Acad. Sci. Lett.* **23**, 113-118 (2000).
10. P. E. John and M. V. Diudea, *Croat. Chem. Acta* 77, 127-132 (2004).
11. M. V. Diudea and A. Graovac, *MATCH Commun. Math. Comput. Chem.* **44**, 93-102 (2001).
12. M. V. Diudea, I. Silaghi-Dumitrescu and B. Parv, *MATCH Commun. Math. Comput. Chem.* **44**, 117-133 (2001).

13. M. V. Diudea and P. E. John, *MATCH Commun. Math. Comput. Chem.* **44**, 103-116 (2001).
14. M. V. Diudea, *Bull. Chem. Soc. Jpn.* **75**, 487-492 (2002).
15. M. V. Diudea, *MATCH Commun. Math. Comput. Chem.* **45**, 109-122 (2002).
16. M. V. Diudea, M. Stefu, B. Parv and P. E. John, *Croat. Chem. Acta*, **77**, 111-115 (2004).
17. A. R. Ashrafi and A. Loghman, *MATCH Commun. Math. Comput. Chem.* **55**, 447-452 (2006).
18. S. Yosefi and A. R. Ashrafi, *MATCH Commun. Math. Comput. Chem.* **56**, 169-178 (2006).
19. S. Yosefi and A. R. Ashrafi, *J. Math. Chem.* 56, 1-9 (2006).
20. A. R. Ashrafi and S. Yosefi, *MATCH Commun. Math. Comput. Chem.* **57**, 403-410 (2007).
21. A. R. Ashrafi and A. Loghman, *ARS Combinatoria,* **80**, 193-199 (2006).
22. A. R. Ashrafi and A. Loghman, *MATCH Commun. Math. Comput. Chem.* **55**, 447-452 (2006).
23. A. R. Ashrafi and F. Rezaei, *MATCH Commun. Math. Comput. Chem.* **57**, 243-250 (2007).
24. A. R. Ashrafi and H. Saati, PI and Szeged Indices of a One-Pentagonal Carbon Nanocone, *J. Comput. Theor. Nanosci.* (in press).
25. B. Manoochehrian, H. Yousefi-Azari and A.R. Ashrafi, *MATCH Commun. Math. Comput. Chem.* **57**, 653 (2007).
26. A. R. Ashrafi and A. Loghman, *MATCH Commun. Math. Comput. Chem.* **55**, 447-452 (2006).
27. A. R. Ashrafi and A. Loghman, *J. Comput. and Theor. Nanosci.* **3**, 378-381 (2006).
28. E. Eliasi and B. Taeri, *MATCH Commun. Math. Comput. Chem.* **56**, 383-402 (2006).
29. A. Heydari and B. Taeri, *MATCH Commun. Math. Comput. Chem.* **57**, 665-676 (2007).
30. A. Heydari and B. Taeri, *J. Comput. Theor. Nano Sci.* **4**, 158-167 (2007).
31. A. Heydari and B. Taeri, Math. *MATCH Commun. Math. Comput. Chem.* **57**, 463-477 (2007).
32. M. Eliasi and B. Taeri, index of zigzag polyhex nanotorus, J. Comput. Theor. Nano Sci. (in press).
33. M. Eliasi and B. Taeri, Hyper Wiener index of zigzag polyhex nanotorus, *Ars. Combinatoria* (in press).
34. A. Heydari and B. Taeri, Hyper Wiener Index of TUC4C8(R) Nanotubes, *J. Comp. Thoer. NanoSci.* (in press).
35. A. Heydari and B. Taeri, *MATCH Commun. Math. Comput. Chem.* **57**, 463-477 (2007) .
36. M. Eliasi and B. *MATCH Commun. Math. Comput. Chem.* **2**, 383-402 (2006).
37. H. Deng and J. Hou, *MATCH Commun. Math. Comput. Chem.* **57**, 503-516 (2007).
38. L. Xu, H. Deng, *MATCH Commun. Math. Comput. Chem.* **57**, 485-502 (2007).
39. H. Deng, *MATCH Commun. Math. Comput. Chem.* **57**, 357-374 (2007).
40. M. Stufa and M. V. Diudea, *MATCH Commun. Math. Comput. Chem.* **50**, 133-144 (2004).
41. A. Iranmanesh, B. Soleimani, *MATCH Commun. Math. Comput. Chem.* **57**, 251-262 (2007) .
42. L. Xu, H. Deng, *MATCH Commun. Math. Comput. Chem.* **57**, 485-502 (2007).
43. H. Deng, *MATCH Commun. Math. Comput. Chem.* **57**, 357-374 (2007).
44. H. Deng, *MATCH Commun. Math. Comput. Chem.* **57**, 461-476 (2007).

Nanotechnology: a Boon or Bane?

B. Bhushan

Petrofac International Limited, Petrofac House, Al Soor Street, POB 23467, Sharjah, UAE
bharat.bhushan@petrofac.ae

Abstract. Nanotechnology is an umbrella term that is used to describe a variety of techniques to fabricate materials and devices on the nanoscale. Nanotechnological techniques include those used for fabrication of nanowires, those used in semiconductor fabrication such as deep ultraviolet lithography, electron beam lithography, focused ion beam machining, nanoimprint lithography, atomic layer deposition, molecular vapor deposition, and further including molecular self-assembly techniques. Applying nanotechnologies raises questions about the environment, health, and societal risks also but that is a huge area of debate. It's too early to say whether Development of Nanotechnology will prove to be a boon or bane for the coming future.

Keywords: risks, environmental issues, health issues, need for regulations, bottom-up & top-down approaches, nanodevices.

INTRODUCTION

Nanotechnology is the Engineering of Functional Systems at the Molecular Scale

Despite the apparent simplicity of this definition, nanotechnology actually implicitly encompasses diverse lines of inquiry. Nanotechnology cuts across many disciplines, including colloidal science, chemistry, applied physics, biology & much more. Two main approaches are used in nanotechnology: one is a "bottom-up" approach where materials and devices are built from molecular components which assemble themselves chemically using principles of molecular recognition; the other being a "top-down" approach where nano-objects are constructed from larger entities without atomic-level control.

Nanotechnology can also be defined as: research and technology development at the atomic, molecular, or macromolecular levels using a length scale of approximately one to one hundred nanometers in any dimension; the creation and use of structures, devices and systems that have novel properties and functions because of their small size and the ability to control or manipulate matter on an atomic scale. Nanotechnology not only will allow making many high-quality products at very low cost, but it will allow making new nanofactories at the same low cost and at the same rapid speed. This unique ability to reproduce its own means of production is why nanotech is said to be an exponential technology. It is a revolutionary, transformative, powerful, and potentially very dangerous—or beneficial—technology. Potential opportunities of nanotechnologies to help address critical international development priorities include improved water purification systems, energy systems, medicine and pharmaceuticals, food production and nutrition, and information and communications technologies.

How Nanotechnology Works

Put simply, nanotechnology is a technology for making things by placing atoms precisely where they are supposed to go. Traditional industrial technologies operate from the top down. Blocks or chunks of raw material are cast, sawed

CP929, *Nanotechnology and Its Applications, First Sharjah International Conference*
edited by Y. I. Salamin, N. M. Hamdan, H. Al-Awadhi, N. M. Jisrawi, and N. Tabet
© 2007 American Institute of Physics 978-0-7354-0439-7/07/$23.00

or machined into precisely formed products by removing unwanted matter. Results of such processes may be rather small (integrated circuits with structures measured in microns, for example) or very large (ocean liners or jumbo jets). However, in all cases matter is being processed in chunks far larger than molecular scale[1]. Rather than being produced through large chunks of material being sawed, planed, and ground to form, most such objects are constructed by tiny molecular machines, such as cells and organelles, working from the bottom up. By organizing individual atoms and molecules into particular configurations, these molecular machines are able to create works of astonishing complexity and size, such as the human brain, a coral reef, or a redwood tree. This approach can produce results that would seem impossible if judged by the standards of conventional top-down production technology, but that are taken for granted in their proper context.

For example, the human body begins as a single cell, a fertilized ovum. Yet a mature human being consists of approximately 75 trillion cells, complexly arranged and of many different varieties[2]. As Eric Drexler states:

> *"Nature shows that molecules can serve as machines because living things work by means of such machinery. Enzymes are molecular machines that make, break, and rearrange the bonds holding other molecules together. Muscles are driven by molecular machines that haul fibers past one another. DNA serves as a data-storage system, transmitting digital instructions to molecular machines, the ribosomes, that manufacture protein molecules. And these protein molecules, in turn, make up most of the molecular machinery".*

Putting these natural molecular machines to work is nothing new, of course, as every living thing does so constantly. Nor is deliberate human programming of those machines particularly new, as it is what genetic engineering (or even selective breeding) is all about[3]. What makes nanotechnology different is that it involves the attempt to go farther than natural mechanisms permit. Using special bacterium-sized "assembler" devices, nanotechnology would permit exact control of molecular structures that are not readily manipulable by organic means (diamond, or heavy metals, for example) on a programmable basis[4]. The key to the application of nanotechnology will be the development of processes that control placement of individual atoms to form products of great complexity at extremely small scale[5]. Efforts in this direction have already generated a substantial amount of theoretical literature, and considerable concrete interest.

What Nanotechnology Can Do

Full-fledged nanotechnology promises nothing less than complete control over the physical structure of matter—the same kind of control over the molecular and structural makeup of physical objects that a word processor provides over the form and content of a text.

Using nanotechnology, production would be carried out by large numbers of tiny devices, operating in parallel, in a fashion similar to the molecular machinery already found in living organisms. Known as "assemblers," these tiny devices would be capable of manipulating individual molecules very rapidly and precisely[6].

The desired molecule would be modeled on a computer screen, the assemblers would be provided with the proper feedstock solutions, and the product would be available in minutes. More complex applications might use groups of assemblers programmed to produce molecules and then hook them together into large structures: rocket engines, computer chips, or whatever is desired[6].

Besides allowing such efficient and powerful manufacturing capabilities, more sophisticated applications of nanotechnology would allow far more subtle applications. For example, specially designed nanodevices, the size of bacteria, might be programmed to destroy arterial plaque, or cancer cells, or to repair cellular damage caused by aging, and then be injected into the body[7]. After performing their tasks, the devices may be induced to self-destruct, or remain in a surveillance mode, or, in some cases, integrate themselves into the body's cells. Such devices would have dramatic implications for the practice of medicine, and for society as a whole.

RISKS OF NANOTECHNOLOGY

Potential risks of nanotechnology can broadly be grouped into three areas:

- The risk to health and environment from nanoparticles and nanomaterials;
- The risk posed by molecular manufacturing (or advanced nanotechnology);
- Societal risks.

Risks from Nanoparticles

The mere presence of nanomaterials (materials that contain nanoparticles) is not in itself a threat. It is only certain aspects that can make them risky, in particular their mobility and their increased reactivity. Because nanoparticles are very different from their everyday counterparts, their adverse effects cannot be derived from the known toxicity of the macro-sized material. This poses significant issues for addressing the health and environmental impact of free nanoparticles.

Health Issues

There are several potential entry routes for nanoparticles into the body. They can be inhaled, swallowed, absorbed through skin or be deliberately injected during medical procedures (or released from implants). How these nanoparticles behave inside the organism is one of the big issues that need to be resolved. The behavior of nanoparticles is a function of their size, shape and surface reactivity with the surrounding tissue. Apart from what happens if non-degradable or slowly degradable nanoparticles accumulate in organs, another concern is their potential interaction with biological processes inside the body: because of their large surface, nanoparticles on exposure to tissue and fluids will immediately adsorb onto their surface some of the macromolecules they encounter. This may, for instance, affect the regulatory mechanisms of enzymes and other proteins.

Environmental Issues

Not enough data exists to know for sure if nanoparticles could have undesirable effects on the environment. Two areas are relevant here: (1) In free form nanoparticles can be released in the air or water during production (or production accidents) or as waste byproduct of production, and ultimately accumulate in the soil, water or plant life. (2) In fixed form, where they are part of a manufactured substance or product, they will ultimately have to be recycled or disposed of as waste.

To properly assess the health hazards of engineered nanoparticles the whole life cycle of these particles needs to be evaluated, including their fabrication, storage and distribution, application and potential abuse, and disposal.

A NEED FOR REGULATION

Regulatory bodies such as the Environmental Protection Agency and the Food and Drug Administration in the U.S. or the Health & Consumer Protection Directorate of the European Commission have started dealing with the potential risks posed by nanoparticles. So far, neither engineered nanoparticles nor the products and materials that contain them are subject to any special regulation regarding production, handling or labeling

Looking at all available data, it must be concluded that current risk assessment methodologies are not suited to the hazards associated with nanoparticles; in particular, existing toxicological and eco-toxicological methods are not up to the task; exposure evaluation (dose) needs to be expressed as quantity of nanoparticles and/or surface area rather than simply mass; equipment for routine detecting and measuring nanoparticles in air, water, or soil is inadequate; and very little is known about the physiological responses to nanoparticles.

A truly precautionary approach to regulation would severely impede development in the field of nanotechnology if we require safety studies for each and every nanoscience application. Consequently, the rush seems to be on to establish a research needs assessment in the nanocommunity to preclude universal safety studies. While the outcome of these studies can form the basis for government and international regulations, a more reasonable approach might be development of a risk matrix that identifies likely culprits.

CONCLUSIONS

The lack of adequate safety knowledge threatens our health, the environment, and the nanotech industry as a whole. Most nanotechnology will likely be harmless, and increased research and funding will create guidelines for utilizing the technology safely, but rapid technological advances with unknown safety risks could damage public opinion. In order to maintain public support, scientists and industry must educate the public about nanotechnology, including realistic descriptions of health risks.

Nanotechnology promises great scientific advances, and it would unfortunate to see that potential compromised by a lack of adequate safety and environmental research. Knowledge of the risks posed by nanotechnology must be improved within the scientific community through increased funding for nanotech-related risks, and in the public through adequate and honest education about the realistic benefits and dangers posed by nanotechnology. To head off future public concerns, both real and imagined, and make full use of the potential of nanotechnology, we need to spend time thinking about the novel risks posed by nanotechnology as well as the novel benefits.

REFERENCES

1. K. Eric Drexler, *Nanotechnology Summary, in* 1990 **Encyclopedia Britannica Science and the Future Yearbook** 162, 163-67 (describing top-down and bottom-up approaches).
2. Arthur Guyton, **Textbook of Medical Physiology** 2 (1986).
3. R. Williamson, *Molecular Biology in Relation to Medical Genetics, in* **Principles and Practice of Medical Genetics** 17-18 (Alan Emery & David Rimoin eds., 1983).
4. K. Eric Drexler, **Nanosystems: Molecular Machinery,Manufacturing, and Computation** 10, 255 (1992).
5. **New Technologies for a Sustainable World, Hearings Before the Subcommittee on Science, Technology, and Space, Committee on Commerce, Science, and Transportation,** 102d Cong. 21 (1992) (testimony of Dr. Eric Drexler) [hereinafter *New Technologies Hearings*].
6. Drexler, **Engines of Creation,** note 17, at 56-63, 57 (describing assemblers).
7. Drexler *et al.*, note 4, at 212-13, 210, 224.

AUTHOR INDEX

A

Abdel Rahim, M., 235
Abdollah-Zadeh, A., 111
Achour, S., 59
Ahmadian, M. T., 82
Ahmed, W. Kh., 38
Al Ahdab, S. K., 32
Al-Dahoudi, N., 100
Al Khawaja, U., 22
Al-Sayah, M., 195
AlSunaidi, A., 43
Aminpour, M., 128
Ansari, I., 143
Arabgari, S., 162
Asadi, H., 89
Ashrafi, A. R., 12, 28
Ashrafi, M. H., 220
Assadi, H., 111
Aziz, Z., 6

B

Babaei, Z., 228
Babanejad, S., 89
Badran, R. I., 201
Benkraouda, M., 22, 123
Bentata, S., 6
Bhushan, B., 250
Boulares, N., 172
Bououdina, M., 67

C

Cherney, E., 209

D

Djelti, R., 6

E

El-Hag, A., 209
Eliasi, M., 243
El-Korashy, A., 235

F

Faiz, M., 147
Fakhraei, B., 89
Farhadi, M., 162

G

Gencer, A., 133
George, K. C., 95
George, T., 95
Ghadiri, M. R., 195
Ghazanfari, N., 133, 138
Ghomashi, M., 19
Guarbous, L., 59
Guergouri, K., 172

H

Habib, A., 157
Hafiz, M. M., 235
Haik, Y., 38
Hajiraissi, R., 216
Hajizadeh, S., 228
Hamadanian, M., 183
Hamed, F., 22, 123
Hamedanian, M., 19
Hasanli, N., 138
Haubner, R., 157
Hosseini, S., 89
Hosseinnia, A., 152
Hussein, A. S., 117
Hussein, M. L., 55

J

Jafari, M., 177
Jahan-Bakhsh, R., 77
Jahanshahi, M., 77, 228
Jakopic, G., 157
Jayaram, S., 209
Joseph, S., 95

V

Vakili-Nezhaad, G. R., 3, 19
Vandadi, O., 183

Y

Yousefi, S., 12
Yousefi-Azari, H., 28

Z

Zahedifar, M., 128
Zakaria, M., 73
Zerkout, S., 59
Ziq, Kh. A., 143

RETURN TO: PHYSICS-ASTRONOMY LIBRARY
351 LeConte Hall   510-642-3122

LOAN PERIOD 1 **1-MONTH**	2	3
4	5	6

ALL BOOKS MAY BE RECALLED AFTER 7 DAYS.
Renewable by telephone.

DUE AS STAMPED BELOW.

This book will be held in PHYSICS LIBRARY until **NOV 3 0 2007**		
JAN 1 0 2008		

FORM NO. DD 22
2M   7-07

UNIVERSITY OF CALIFORNIA, BERKELEY
Berkeley, California 94720–6000